NUMBER 435

THE ENGLISH EXPERIENCE

ITS RECORD IN EARLY PRINTED BOOKS
PUBLISHED IN FACSIMILE

The publishers acknowledge their gratitude
to the Curators of the Bodleian Library, Oxford
for their permission to reproduce the
Library's copy, Shelfmark: 4°.A.4 Art BS (4)
and the pages B4v, C1r, C3v, C4, D1r, from
Shelfmark: Savile C.c.9 and the two maps
from Shelfmark: Ashm.417

Library of Congress Catalog Card Number:
78-38156

S.T.C.No. 3117 + 3120

Collation: A-H^4, I^2 + 2 fold. maps

Published in 1972 by

Theatrvm Orbis Terrarvm Ltd.,
O.Z.Voorburgwal 85, Amsterdam

&

Da Capo Press Inc.
-a subsidiary of Plenum Publishing Corporation-
277 West 17th Street, New York N.Y. 10011

Printed in the Netherlands
ISBN 90 221 0435 4

Astrolabium Uranicum Generale.

A Neceſſary and Pleaſaunt
ſolace and recreation for Nauigators
in their long Iorneying, Containing the vſe of
an Inſtrument or generall Aſtrolabe : Newly for
them deuiſed by the Author, to bring them
skilfully acquainted with all the Planets, Starres, and
conſtellacions of the Heauens : and their
courſes, mouings, and apparences,
called the (Vranicall Aſtrolabe.)

In which,

Agreeable to the Hipotheſis of Nicolaus Copernicus, the Starry Firmament is appointed, perpetually fixed,
and the earth and his Horizons continually mouing from
Weſt towards the Eaſt once about euery 24 houres.

Fraught alſo by new deuiſe with all ſuch neceſ-
ſary ſupplements for Iudiciall Aſtrology, as ALKA-
BITIVS & CLAVDIVS DARIOTTVS
haue deliuered by their Tables.

Wherevnto for their further delight he hath anexed another inuention, expreſsing in one face the whole Globe
terreſtriall ; with the two great engliſh voyages
lately performed round about the world.

Compyled by IOHN BLAGRAVE of Reading
Gentleman, the ſame wellwiller to the
Mathematicks. Anno. 1596.

Printed by Thomas Purfoot, for William Matts.

¶ To the Right Honorable Lord, the Lord Charles Haward, Baron of Effingham, Lord high Admirall of England, &c.

Right Honorable, for so much as my selfe, with the whole progeny of our name, haue beene alwayes exceedingly bounde, Next to the Right Honorable our good Lord the Lord Burleigh, Lord high Treasurer of England, Vnto the late right Honorable Sir Fraunces Knolles the elder Knight, Treasurer of her highnesse most honourable houshold, deceased, in regard of their sinceare iustice, and charitable respect of most iniurious and wicked practises heretofore vehemently prosecuted against vs, for no lesse then the liuing we all possessed. Both whome therefore (vnder your Lordships good fauour) I can not, but in all my best indeuours, during life remember, Not forgetting as in duty I am bound, the Right Honorable Sir William Knolles Knight, now Controller of her Maiesties sayd most honorable houshold, and such other his right worshipfull

THE EPISTLE DEDICATORY.

full and honorable minded off-spring, as still remaine our good fauorers. Amongst whom, the right Worship. M. Robert Knolles more then a yeare passed of his accustomed kindnes aduertised me, that your L. I know not by what meanes, vnderstanding of my indeuors in these Mathematick Sciences, was pleased to take further notice of me by my personall presence. For the which, as the time hath not bin fitting by reason of your L. most serious important & honorable imployments, aswell to the seas as otherwise this yeare 1596 well knowne to the whole world. So haue I bin willing to catch hold of the opportunity thereof so far: as in the meane time to hammer out (if selfliking deceiue me not) a rare piece of Mathematicke stuffe, to bring vnto your L. view, together with my self: tending both to the profit and pleasure of all worthy english Gent. that either addict themselues to contemplacion at home, or to Marine traueyles abroad. Which if it may receiue fauorable allowance at your good L. hands, I shall thinke my selfe suffiicietly defended against all ignorant or malignat detractors, and be the rather hereafter encouraged to go forwards with like matters. Thus praying vnto God for the long preseruacion of your Honour, I cease.

Your Lordships most humble, Iohn Blagraue.

To the curteous Reader.

Entle Reader, the eares of my guilty conceit being continually troubled with such a noyse and clamor for my (Organum Vranicum,) and the second part of my Mathematicall Iewell, so long and largely promised, as the greedy Auditors are woont to make at a Stage-play: calling, Come away howe, with boyes throwing vollyes of stones ratling at the gates of my otherwise busied imployments, for staying so long: Maketh me heere in haste (God graunt to your liking) to come vp the Stage, with a little, but rare piece of Mathematicke Musickly pleasure, not impertinent to the matter to be acted: to stay your languishing expectations, both woorth the acceptacion (which I dare boldly aduouch against the doggedst detractor that shall contradict) and sufficient to entertaine the time vntill Midsomer or Michelmas Terme at the farthest, when as (if God permit) the stately Comedy of Queene Vranya hir Pageants, shall abundantly offer themselues to your pleasant viewe, Sollace your selues I pray you heerewith the whiles, and be bold with me if in any thing you doubt. You shall haue me God willing once euery Terme in London ready to your wish, at a poore lodging within Maister Greenes Wharfe neere vnto Charing Crosse: and at Maister Ralfe Iacksons at the signe of the Swan in S. Paules Churchyard, or at Maister William Matts Stationer, at the signe of the Plough, ouer against S. Dunstons Church in Fleetestreete, who hath the Impression of this Booke, you shall euer know whether I be at any time in the Citty: With whome, I haue taken order to furnish with these Instruments and their supplements any that shall want them.

The Contents of the Chapters of this worke, of the Vranicall Astrolabe.

Cap. 1. OF the partes of this new Astrolabe.

Cap. 2. Of the Celestiall proiectinent in this new Astrolabe, and how the longitudes and latitudes of the Starres are thereby had.

Cap. 3. Of certaine abbreuiated tearmes or phrases of arte appertayning to this new Astrolabe.

Cap. 4. How to place Carects of the Planets in this Astrolabe to serue there the whole weeke after.

Cap. 5. How by this new Astrolabe to finde the true place of the ☉ perpetually.

Cap. 6. How to rectifye the true place of the Sunne to all places of the world.

Cap. 7. How to rectifye the place of the ☽ or of any other Planet, to any other Meridian.

Cap. 8. How to know the declinacion and right ascenscion of any Planet, fixed Starre, or point of the heauens, and their degree of Culminacion.

Cap. 9 How to finde the true Meridian in any Countrey by helpe of this Astrolabe.

Cap. 10. How to attaine the latitude or Poles Eleuacion in any countrey vnknowne, and to choose out the Verticall point in the noone-line, the Horizontall point in the North line, and the Zenith line seruing thereunto.

Cap. 11. How to take the Meridian altitude of the Sunne or any Planet or Starre, and thereby to get the latitude in any Countrey.

Cap. 12. How to know what day of the yeare it were if it were forgotten.

Cap. 13. How to know what it is a clocke at any time either night or day, by the Sunne, Planets, or Starres.

Cap. 14. Another more easy way to finde the houre by the Starres.

Cap. 15. Another easy way by the ☽.

Cap.

The Contents of this Booke.

Cap. 16. How to know the Meridian altitude of the ☉ or any other Planet, Starre, or point of the heauens in any Countrey, and what houre of the day or night any of them shall culminate or come to the Meridian of that place.

Cap. 17. How to know the houre of Sunne-rising and setting, and the length of the day and night, together with his oblique ascension, and his difference of ascenscion any day in the yeare.

Cap. 18. To know the houre of rising or setting of any other Planet, Starre, or point in the heauens, and their diurnall and nocturnall arches, and their oblique ascenscions and difference of ascenscions in any countrey.

Cap. 19. How to set particular degrees on the Almicantiser for any perticular latitude.

Cap. 20. How to know the Amplitude of rising or setting of the Sunne or Starres.

Cap. 21. To know what Starres do neuer rise or set in any latitude, and which do once a day touch the Zenith point, or Horizontall point.

Cap. 22. To know any day and houre what Starres or points in the heauens are in or neere the Meridian, or do rise or set in any latitude.

Cap. 23. How to know the houre, distance of any Planet, Starre, or point of the heauens from the Meridian, any day and yeare proposed in any latitude.

Cap. 24 How to know the height of the Sunne or of any other Planet or Starre for any houre of the day any yeare in any latitude.

Cap. 25. How you shall most easily learne to know all the constellacions and Starres in the element it selfe.

Cap. 26. Of the 4 Cardines, and the sundry rising and setting of the fixed Starres with the Planets.

Cap. 27. Of the 12 houses, and setting the Figure.

Cap. 28 Of the aspects of the Planets and Starres, and how to finde them.

Cap.

The Contents of this Booke.

Cap. 29. A new deuife of the Author to get the Planetary houres moſt eaſily in all latitudes, and of their Lordes regnant.

Cap. 30. How you ſhall know to ſet forwards and backwards any of the fixed Starres vnto their true places for any ages paſt, or to come.

Cap. 31. How by helpe of this Aſtrolabe to get the longitude and latitude of any Planet, yea, or of any Starre or Comet ſeene.

Cap. 32. A ready way by helpe of this Aſtrolabe heereafter to obſerue the motions of Comets.

Astrolabium Uranicum Generale.

Heere followeth the booke of the vse of the *Vranicall Astrolabe*.
Newly deuised by Iohn Blagraue of Reading, Gentleman.

Cap. 1.

¶ Of the partes of this new Astrolabe.

His Astrolabe hath three generall partes, that is to say, the Celestiall, the Zenitfer with his Cursor, and the Almicantifer with his Pointer.

I call that part the Celestiall heere, which in other Astrolabes is called the Mater, of which I will largely speake in the next Chap.

The Zenitfer beeing but as a playne rule of one ynch bredth, mooueth vpon the Celestiall in stead of a Reete vsed in other Astrolabes: and hath a circular Lymbe thereto annexed neere vnto the Centre, which I call the Circulus horarius, or houre circle, with the 24 houres, and their partes thereon, as you see. Neither is it materiall of what quantity that houre circle or lymbe be of, so it be concentricke to the Astrolabe, and placed so, that it hide fewest Starres in his moouing. And for seruice of this circle, there is written at the vpper end of this Zenitfer Meridies, and at

B the

the lower end Media nox, and therefore the fiduciall line thereof from the Centre vpwards to Meridies, is generally called the Noone line, the other part from the Centre downewardes vnto Media nox the North line. Then hath it 90 Zenith lines standing in ascents one aboue another like Organ pipes, in stead of the 90 generall Horizons before time vsed, issuing or growing out of the 90 Zenith points: for the lowest points of these 90 lines neerest vnto the centre vnto which their numbers are set from 0 to 90, do represent the true Zenithes of all places. These 90. Zenith lines, are by certaine thwart curued lines parted euery one into 90 vneuen parts, of which some run out infinitely, and those partes I call the Altitude rootes, or the Almicantare rootes, and for breuities sake sometimes Rootes without any addition. These Rootes do helpe vs to generall Almicantares, or circles of Altitude for any Latitude by a rare deuise by me first found out. They are numbered on the outer side of the Zenitfer from their extreame points inwards towards their sayd Zenith points, ending at them with 90. Also to the Fiduciall line of this Zenitfer is anexed a scale, in which are planted the Radiall degrees of the proiectement on both sides of the Scale, because they require double numbering for sundry purposes. And of those which bound on the Fiduciall line, those at the noone line are numbered from the Equinoctiall intersection both waies, vz. towards Meridies, as farre as the Zenithfer will giue leaue, which are about 36, and towards the Centre, ending there at 90. which 90 radiall degrees of the noone line towards the Centre do serue (with the other 36) to descry the declinations of the fixed Starres and Planets, but chiefely they serue for the 90 Zenithes which appertaine to the 90 generall Horizons, and for distinction sake I call them the 90 Verticall points, and they exactly answere vnto the 90. lower ends of the 90 Zenith lines, which before I called the 90 Zenith points, in so much that the fiduciall line of the Cursor in his moouing, cutteth the Zenith point, and the Verticall point

of

Astrolabe.

of any latitude both at once. The other Radiall degrees in the North line, are numbred from the centre towards Media nox, ending at the Equinoctials interserion therewith at 90, and do serue for the 90 intersections of the 90 generall Horizons therewith, and therefore I call them the 90 Horizontall points. And of those radyall degrees which bound on the other side of the Scale, which indeede serue to the getting of the 12 houses generally, as is shewed in the 27 Cap. The Norther 90 towards Media nox are numbred inwards towards the Centre, ending there at 90 contrary to the former. And the Souther of those radiall degrees are numbred from the Centre outwards towards Meridies two to one, as farre as they will go, (that is to say) euery 10 make but 5, and these do serue but in stead of rootes to the other, and therefore I call them the Domifieng rootes. Also there adioyneth vnto the Radiall scale of the North line betweene the centre and the Zenith lines a certaine scale of slope lines numbered in the middest, beginning but from 60 and ending at the centre with 90, which I call the Domifieng Scale, for that the two extreames of euery of those slope lines, do help to get 8 of the 12 houses in that latitude, whose number it sheweth, as in the 27 Cap. is manifested: and vnto the outer edge of this Domifieng scale, are also anexed part of the former radiall degrees, all hauing relacion to them on the North line. Then hath this Zenitser a certaine runner or Cursor, whose fiduciall line alwayes runneth exactly square to the fiduciall line of the Zenitser. And at the common meeting of both those fiduciall lines, this Cursor hath iointed vnto him a kinde of Labell or loose scale, which I call the Almicantiser, who to supply the generall vse of the Almicantares, hath no manner of matter on him but another Cursor, which I call a Pointer.

But now for any man that desireth to haue this Astrolabe more particularly commodious for one latitude, vz. for the countrey or place of his abode, then shall he haue a kinde

The Vranicall

of Reete or moouer made, carying a circle or limbe of houres of equall diametre to the Celestiall, and within it the Horizon circle of that place particularly described after Stophlerus, and graduated into his 360 prospectiue degrees, as in the fourth booke 4. Cap. of my Iewell is taught, together with the arches distinguishing the 12 houses, as in Stoph. lib. 1. propo. 6. is taught, writing to each house his number and significations. And vnto the noone-line furnished with the former radiall degrees, numbred from the Horizon circle, into twise 90. both ending at the Verticall point, you shall neede to annexe but only one of the 90 Zenith lines, vz. that which serueth to your latitude and his rootes, with the Cursor: Neither neede your Almicantifer haue his sliding pointer, but in stead thereof, the particuler degrees of altitude seruing to that latitude, which are compounded of his 90 Zenith rootes, and the 90 radiall degrees. in the North line, as in the 19 Cap. is taught, all which shall better appeare by such Astrolabes as I meane to set abroade ready furnished.

Cap. 2.

¶ Of the Celestiall proiectment in this new Astrolabe, and how the longitudes and latitudes of the Starres are thereby had.

Needlesse it is to stand vpon the Fabrication, or long particularizing of this Instrument or his partes, since they are but the Copernician representation of the constellations fixed, and the 90 generall Horizons moouing.

The Fixed part which in other Astrolabes is called the Mater, I call for breuities sake the Celestiall, For that in it I haue prospectiuely

Astrolabe.

prospectiuely proiected the Globe Celestiall, that is to say, the Globe of the fixed Starres, with the Zodiack, and his North Pole (marked with D) the Images of each constellation, and the Starres appertaining, in such sort as in the Globe they are vsuall, with each fift circle of Longitude and Latitude of the Zodiacke, Euery fixe of those fiues of the Longitude including a signe in the heauens, being 12 in the whole, which 12 are distinguished with twisted lines somewhat bigger from the rest, and vnto them neere vnto the Limbe are written the names of the 12 Signes as they passe in order, Aries, Taurus, &c. together with their natures and quallities. Also, those three Signes of the watry triplicitie are filled vp with waues, and the three fiery with flames, and the three earthy with bigge prickes, and the three ayry left white. Then are there seuen seuerall magnitudes of Starres, each bigger and brighter then other, whose fashions are set downe with their numbers in the Signe ♊ before the brest of Orion, by which they are easily heerein descryed. The excentricke circle Orbs or Lymbe, lying wholy within it, is the Zodiacke deuided into his 12 Signes by the foresayd 12 mayster circles of longitude, and euery Signe into his 30 prospectiue degrees. Also the degrees are made with diuers distinguishments, for the Gradus masculini, feminini, lucidi, tenebrosi, vacui, fumosi, puteales, fortunam augentes, & debilitantes, all which are sufficiently to be descryed by the two Directoria Zodiaci there placed. Then is there a little within the Zodiacke a rowe of Carecters of the Planets placed at euery 10 circle of longitude, signifying the faces of the Planets, and likewise another rowe without the Zodiacke at the beginning of the termes of the Planets, euery face and terme continuing vnto the carect next him in sequens of the Zodiacke, all which will better appeare, if you ioyne vnto this booke Claudius Dariot his introduction to the iudgement of the Starres, which by good hap is now newly imprinted in english together with this, and a second part therewith,

with, which at the first impression was omitted. And specially you must note and remember well, that the very outermost circle of this Zodiacke, is the clipticke circle, within which is contayned the North-halfe or Hemisphere of all the Starres and Constellacions in the heauens, and without it so many of the South Starres and Constellacions, as the proiectment would admit. Then is there also lying within this Astrolabe a blacke concentricke circle, representing the Æquinoctiall circle, and proiector of this worke, but without any diuisions, because they are better supplyed by the limbe. Also there is a scale of the radiall degrees, whereon this proiectement is made, placed at the master circle of longit. beginning the Signe ♋ numbred from the Equinoctiall both wayes, which Equinoctiall cutteth the sayde eclipticke circle in the pointes B and C beginning ♈ and ♎, and those points are called the two Æquinoctiall Intersectiones which do moove vppon the poles of the Zodiacke backwards, or in precedentia, of the Signes euery 67 yeares almost a degree equally : carying about with them all the foresayde circles of Longitude and Latitude, leauing the Starres stocke still in their places, By which meanes it commeth to passe, that although the fixed Starres haue no manner of motion at all, yet they seeme from time to time to haue gone as is sayd euery 67 yeares almost a degree forwards, as the aunctient Astronomers thought : where as in very deede (as Copernicus sheweth) the Zodiacke and all his circles do slide so much backwards from them: And by reason this sliding backe of the Æquinoctiall points, is as I haue sayd performed on the poles D and his opposite of the clipticke, therefore it commeth to passe that the fixed Starres could thereby neuer change theyr Latitudes, but keepe the same Latitudes as euer they did. This sliding backe, is called Precessio Æquinoctiorum, of which wee will intreate more in the 30 Cap. And now will returne to this proiectement : In which it is very easye to imagine euery

square

Astrolabe.

square into fiue equall partes, and so thereby to knowe the Longitude and Latitude of any Starre therein sufficient for any common vse. As for example, the great Starre in the left foote of Orion, placed in the Signe ♊, if you imagin the spaces betweene each fiue circles of Longitude and Latitude deuided into fiue partes, then shall you conceaue that the middle pricke of this Starre is heere in longitude about 9 degrees 30 minut. of ♊, and in latitude South from the clipticke circle about some 31 degrees 30 min. and so for any of the rest.

Cap. 3.

¶ Of certaine abbreuiated termes or phrases of arte, appertayning to this new Astrolabe.

Ecause this Astrolabe is newly deuised and his tackling, much differing from all other heeretofore both in fashion, vse, and names, I haue therefore heere thought good to premise certaine abbreuiated termes thereunto fitly appertaining, least otherwise such long iteracion of diuers circumstances as at euery time I mention them I should be driuen to vse, should make my writings yrkesome, and my booke ouer long. Which tediousnes to auoyde, I haue further not onely couched diuers precepts and propositions almost into euery Chapter, but also incorporated the examples with them, yet in another letter, to the end you may if you list first reade ouer euery Chapter, leauing out the example euen as though it were not. And if you vnderstand it so, it is well, if not, then reade it, and practise the example withall.

The

The Vranicall

The Verticall point,
The Zenith point,
} Are all one in signification, but in this practise the one is taken in the noone line, the other in the Zenith line of the place, as in the 1. Cap. is noted.

The Horizontall point, is alwayes that radiall degree of the north line, reckned from the centre which is equall to the latitude, or poles elevation, explaned both in the 1. & 10. Cap.

The Zenith line, explaned in the 1. and 10. Cap.

The North and South degrees, are the 90. radiall degrees on the noone and north lines, on each side of the Verticall point explaned in the 10. Cap.

The East and West Hemispheares are thus, in every constitucion of this Astrolabe, the Fiduciall line of the Zenitfer deuideth the Celestiall, in halfe of which, that halfe Eastwards of the noone line towards vi. of the clocke in the morning, I call the East Hemisphere, the other halfe I call the West Hemispheare.

To play the Almicantifer in the East or West Hemisphere is as much to say, as to keepe him moouing wholy on the East, or West side of the noone line accordingly.

To apply the pointer to any Starre or Carect, is to leade the Zenitfer, and the Almicantifer about, till you may cause the Apex of the pointer to touch the carect, or the middle prick of the Starre.

To labell the Hower to the Sunne or Starres, or them to the hower.
To prepare the Zenitfer to any altitude.
} explaned in the 13. Cap.

To prepare the Zenitfer to the Horizon, is explaned in the 17. Cap.

To prepare the Almicantifer to a particular latitude, explaned in the 19. Cap.

Also when it is sayd, Place the Zenitfer, or Labell to any Planet, Starre, or point: it is meant, that you must place the fiduciall lines thereunto, and that with relation to the centre.

Astrolabe.

The complement of any degree is so much as it lacketh of 90. degrees.

Lastly, you shall vnderstand, that in writing of this whole worke, my meaning was that euery man should haue the furniture of this Astrolabe (whether generall or particular) made of mettall, as commonly the furnitures of Globes are. For which, if any man shall repaire or send vnto me either by him selfe, or by Maister Matts the bookeseller, dwelling at the signe of the Plough ouer against S. Dunstons Church in Fleetstreete, who shall easily at all times send any letter or notice weekely vnto me by our Carryers, I will take such order that he shall haue them as reasonably done as may be. And to the ende that euery man may see how it ought to be made in mettall, I haue caused a rude patterne of each furniture in mettall to remaine with the sayd Maister Matts. Notwithstanding, for such as will not go to that little charge, I haue vpon better aduisement caused the Zenifer and Cursor to be imprinted with the Astrolabe, to be set either on pastboord, or cuttlers scale to serue the turne: But in stead of the Almicantifer and his pointer, then must they vse a thred and pearle, which shall do in a manner as well, for the pearle shall in all actions supply the steede of the pointer. And in steede of the ledge of brasse in the beginning of the next Cap. mentioned, he may haply cause a breadth of lantornes horne to be riuetted ouer the Zodiacke, vnder which to place the Carects of the Planets, made of paste-boord, quilles, or such like.

Cap. 4.

¶ How to place Carects of the Planets in this Astrolabe, to serue there the whole weeke after.

OR this purpose you shall take a paire of Compasse, and pitch his one foote in the line beginning ♄ at 25 degrees from the-cliptickes pole, and extend the other in the same line vnto 7 degrees

The Vranicall

grees of latitude, and with that quantity describe a blinde circle, on which blinde circle so made you shal place a narrow hoope of lattin brasse, riuetted on the Mater with 4 or 8 riuets at the most, with so many thin bonsters vnder euery riuet, that the hoope presse not hard on the Mater, to the end that you may thrust vnder it some 9 Carrects, 5 of them carying two Apices or pointers ioynted together, made all of thin mettall and very narrow, vz. 7 for the 7 Planets, and 2 for the Moones cap, and Cau. draconis, there to sticke very fast till they be weekely remooued, and let one of the Apices of each Carect be lesse and sleighter then the other, which you shall better vnderstand by such as I haue, and shall cause to be ready made to your hand then by manifold words.

These things prouided for, then at the beginning of any weeke of any yeare proposed (Admit for the beginning of the first Sunday in March, which shall be Anno Domini 1598.) you shall then by my Organum Vranicum, which God willing, long before that day shall be extant, or by your Ephemeris, or other Tables of the Planets motions get the true places of euery of the 7 Planets as well in Longitude as Latitude, and of the ☊ and ☋, (which on the said Sunday being the 5 day of March 1598 you shall by Stadius his Ephemerides, finde the ☉ to be 25 degrees 3 minutes in ♓, the ☾ 2 degrees 40 minutes in ♋, hir Latitude then 4 degrees 30 min. North, ♄ 2 degr. 32 min in ♎, in Latit. 2 degr. 44 min. North, ♃ 9 degr 26 min in ♊, in latit South 0 38 min. ♂ in ♐, 1 degr. 39 min. in ♋, in lat. 2 degr. 13 min. North, ♀ 10 degr. 51 min. in ♉, in latit. 1 degr. 40 min. North, ☿ 4 degr. 26 min. in ♈, in latit. 2 degr. 58 min. North ☊ 6. degr. 16 min. in ♓, the ☋ is alwaies opposite, vz. 6 degr. 16 min. in ♍, both which, and likewise the ☉ neuer haue latitude. Which places you shall reddily finde out in this Astrolabe, by meanes of the circles of Longit. and Latit. thereon described, and vnto them set the chiefe Apex of each Planets Carect accordingly, carefully prouiding that you set the Apex of the ☉ his Carect

iustly

Astrolabe.

iustly to the-clipticke circle mentioned in the 2 Cap. at his exact degree and minute, because he hath no Latitude, and he is the Chiefetaine guide in all our worke following, and so likewise the ☊ and ☋, as for all the rest, you shall place their apices vnto the paralell circle, expressing his Latitude at his degree of Longitude reckoned in the same paralell by help of the circles of longitude: Which done, you shall then againe either by your Vranicum or Ephemerides, get in like sort the places of them all for the beginning of the next Sunday following (which will be the 12 day of March that yeare 1598. the ☉ then being 1 degr. 57 min. in ♈, the ☾ 27 deg. 8 min. in ♍, hir latit. 1 degr. 47 min. North, ♄ 1 degr. 56 min. in ♎, ♃ 10 degr. 24 min. in ♊, ♂ 4 degr. 10 min. in ♋, ♀ 16 degr. 46 min. in ♉, ☿ 28 degr. 32 min. in ♓, ☊ 5 degr. 54 min. in ♓, as for their latitudes they alter little or nothing in 6 or 7 dayes.) and to those places accordingly sought in this Astrolabe, you shall set the petty Apices of each Carect. There shall they serue you for all that weeke to singuler effects, and presently you may note thus much, that if the petty Apex happen to be thus placed in sequens of the Zodiacke from the chiefe Apex, then be sure the Planet is direct, if in precedens, (as in ☿) then is he Retrograde, if iust one vppon another, then Stationary. And full easy will it be to imagin the spaces betweene the two Apices of each Planets carect, into seuen equall partes for the seuen dayes of the weeke included betweene them, as in the 2. Cap. you imagined the Squares into fiue partes, and so reddily to conceiue where each Planet is any of those seuen dayes, which in none of them can exceede twise 7 Degrees, (the Moone excepted) whose gate in common accoumpt is euery two houres one whole Degree, and euery foure and twenty houres twelue Degrees. Wherefore for her, epther you must relye vppon your Almanacke, or haue foure single Carects at the least to be set two dayes in sunder apeece, that is to saye, the one to bee set at the Moones place

The Vranicall

for Sunday noone (for Astronomers alwayes begin their day at noone, because the Meridian is Horizon rectus) the second to be set for Tewsday noone, the third for Thursday noone, and the fourth for Saterday noone, and so by adding or subducting of 12 degrees for a day, and two degrees for an houre, her place shall sufficiently well be had. Her motion is very swift, and her Prosthapheresis, or Æquacions, caused by her epicicles, may happily breede error of 7 or 8 degrees in one weeke, from that common accoumpt of two houres to a degree, otherwise two Carects might haue sufficed, but in two dayes her greatest gate being but 12 degrees either before or behinde one of the carects, can breede no error to be regarded, if you recken it from that carect vnto which she is neerest.

Cap. 5.

¶ How by this new Astrolabe to finde the true place of the Sunne perpetually.

IN my Organum Vranicum shortly like to come foorth, you shall haue the Kalender of the yere, the Cicle of the Sunne, and all such necessary supplements. And therefore in this Astrolabe, which I do appoint to be alwayes anexed thereunto, and to be but the back side, or Dorsum Vranici, I haue framed no Kalender at all or Theoricke of the ☉ as commonly all other Astrolabes haue. Yet to serue the turne vntill the Vranicum come foorth, or for them that would content themselues without the other, as happily some will in regard of the charge of the other, which can not well be framed but of mettall, I haue in the 12 Signes of the Zodiackes Orbs placed the 12 letters of the 12 Moneths, vz. I for Ianuary, F for February, &c. at such degree and minute of the eclipticke as they doe make their entraunce the Radicall yeare of my sayd Vranicum, being the

yeare

A Strolabe.

yeare after Chrift 1600. by help of which you fhall for euer if you will, finde the true place of the Sunne within 3 or 4 minutes, so you regard but three obseruaunces which I shall heere deliuer you. The firft is, concerning the Leape yeare, and is remedied by rebating some quarter of a degree for euery yeare following the Leape yeare till it be Leape yeare againe. The second is, concerning the fiue odde dayes in the yeare aboue 360. and is remedied by rebating for euery day of the moneth, proposed but one minute. The third is, concerning ages paft or to come, the remedie whereof, I will shew in the end of this Cap. Now therefore for the time present, vz. for some 20 yeares if neede be before or behinde the yeare 1600. it serueth fitly as it is vpon euery Leape yeare, (because his roote 1600 is a Leape yeare) without any obseruacion more then for the fiue odde dayes which is generall in euery yeare : and for any common yeare with very little trouble, as I shall shewe in this manner : Wherefore if you will know what degree and minute the Sunne is in vpon any day and yeare assigned. First, seeke by the Zodiacke for the letter of the Moneth in which your day is (vz. For the letter I seruing for Iune, which is placed at the 20 degree and somewhat more in ♊, and the number 6 ascribed thereto, signifieth 6 minutes more then the 20 degrees, the exact place of that letter I.) And from that letter (vz. I.) number forwards so many degrees of the Clipticke as your day (the 8 of Iune) proposed is within your moneth (Iune) proposed, rebating so many minutes, according to the second observaunce before mentioned (vz. for your 8 dayes, recken 8 degrees in sequens of that 20 deg. 6. min. of ♊ lacking 8 min.) and where that accoumpt falleth out (which you shall finde vppon that reckning and rebating of 8 minutes, to be 27 degrees 58 minutes of ♊,) there shall be the true place of the ☉ if it be a Leape yeare (as this yeare 1596 is.) and there (vz. at the saide 27 degrees 58 minutes of ♊) you may place the Carect of the Sunne for that day, as in the last Chapter is taught.

C 3

The Vranicall

taught. But if it shall be the first yeare after the Leape yeare, then shall you rebate a quarter of a degree, or $14\frac{1}{4}$ min. (if you will be precise) if the second yeare, then rebate halfe a degree, if the third, rebate $\frac{3}{4}$ of a degree, and then euery Leape yeare it returneth to his owne place againe. And this for almost twenty yeares eyther before or behinde the Radicall yeare 1600 shall serue passing well and reddye within some foure or fiue minutes, but for any time after the yeare 1600 you shall neede but adde, and for any time before subtract some halfe quarter of a Degree (or $8\frac{1}{4}$ minutes if you will be precise) for euery 20 yeares. (As for Example for the place of the ☉ the 8 day of March in the yeare 1700, becauſe betweene the yeare 1600 and 1700, there are fiue times 20 yeares, therefore I adde fiue diquarters of a degree (or to be precise fiue times $8\frac{1}{4}$ minutes which maketh $44\frac{2}{4}$ minutes) vnto the foreſaid 27 degr. 58 min. in ♊, and it maketh 28 degrees, $42\frac{2}{4}$ minutes, which is the true place of the ☉ on the 8 day of Iune, Anno. 1700.) And becauſe you shall worke the more certaine, I haue to euery of the 12 letters of the moneths, ſet the number of the minutes of that degree where his right place ſhould be, which place is alſo deſcryed by a little ſtroke or ſtring cutting the true place of the ſame Degree, and leading thence to the letter, if it be not placed directly therewith.

Alſo you muſt not forget for ages paſt or to come, to adde or ſubtract the Æquacion of the 8 Speere, as in the 30 Cap. ſhall be ſhewed according to the words Adiectiue or Ablatiue there mentioned, for that will from time to time alter this true place of the Sunne, euen vnto $1\frac{1}{2}$ Degrees, about 800 yeares to come, or paſſed. Although ſome others haue ouerſlipped in their Theoricks of the ☉ both this, and the third obſeruacion before. (For example you ſhall by the 30 Cap. finde the 8 Speares Æquation $12\frac{1}{4}$ minutes ablatiue for the yeare 1700, and ſo will the true place then be but 28 degrees $30\frac{1}{2}$ minutes for the 8 of Iune.)

After

Astrolabe.

After many words thus spent, the precept is briefely thus: Suppose every degree of the-clipticke to be a day of the yeare, rebating for every day a minute, and for every of the three common yeares 14¼ minutes, and thus farre it serueth for all times. But now if your time be after 1600, you must for the motus equalis compositus adde 8¼ degrees for every 20 yeares, and for every 20) take before 1650 adde 5 min. for the 8 spheeres equation: and contrariwise, if being before 1600 and after 1650 you must subduct as much. Note, that I haue been sithence enformed that our honourable Lords departed the 3 of June, but let the 8 of June now stand for the example, nothing hurtfull.

Cap. 6.

¶ How to rectifye the true place of the Sunne to all places of the world.

MY good meaning determining this Astrolabe as a fit recreation or play-game for trauellers by Sea, to driue away the tediousnes of their iourneying, may soonest of all deceiue them in getting the true place of the ☉, whether by the Ephemeris, or by the last Cap. or by my Organum Vranicum, if they haue not a speciall regard to the changing of their Meridians as they trauaple: for which cause, I haue caused a new description of the terrestriall Globe to be made after my fashion in one face, where others make them still in two faces: and both that terrestriall and this celestiall, are principally meant by me to serue vnto my Vranicum, and to be placed in the inner sides of his case as necessary appendices, though of itselfe it bee an Instrument sufficient without these. Of this terrestriall Iooe at this time (although hauing manye extraordinary conceites to applye him for Nauigation, and other necessities to write of) determine to deliuer but this Chapter, therein to let you vnderstande,

that

The Vranicall

that in the limbe circle about it which representeth the Æquinoctiall circle at the Meridian cutting vpon London, I haue placed a rundle with a crosse, and at the opposite Meridian the number of 12. signifying 12 houres, and betweene these on both sides are 11 other rundles for houres, so is that whole circle deuided into twise 12 houres, and numbred both wayes from the crosse rundle, signifying how many houres the Sunne or any Planet or Starre hath to go from our Meridian at London (when any of them come vnto it) before it can come to the Meridian of another place, or being in another place, how many houres it must go ere it come to ours. And in regard that the gate of the ☉ is euery 24 houres a degree of the-clipticke, therefore you must adde or subtract from his place found either in the last Cap. or the Ephemerides, or my Vranicum, according to the rate thereof, vz. for euery 12 houres, about 30 minutes of a degree, and for 6 houres 15 minutes, &c.

As for example, admit you being a trauayler should be the fift day of March 1598 in the 4.Cap. mentioned in the great City of Quinsay in the countrey of China, and there seeking for the true place of the Sunne at noone, had found it 25 degrees 3 minutes in ♓ as in the 4.Cap. it was. Now because the Meridian of Quinsay appeareth in this Terestriall by the said houre rundles almost 9 houres distant from the Meridian of London or rather of Anwerp if you worke by Stadius Ephemerides, or the last Cap. and that towards the East: Therefore the ☉ commeth so much sooner to the meridian or noone at Quinsay, for which cause you must in this case rebate so much of the Sunnes diurnall gate of 60 minutes, as 9 houres commeth to, vz. 22 minutes, so will the Sunne that day at noone in that countrye be but 24 degrees 41 min. in ♓, which in the 4 Cap. was found to be 25 degrees 3 minutes in ♓. And if your place had beene so much Westwards, then should you haue added 22 minutes vnto 25 degrees 3 minutes in ♓. You may also performe the same

reddily

Astrolabe.

reddily by the next Cap. applying the Labell vnto 30 minut. of the sextans, for so much is the semidiurnall gate of the Sunne commonly.

Cap. 7.

¶ How to rectifye the place of the ☾ or of any other Planet to any other Meridian.

Here is more neede to haue regard of this matter for the ☾, because her motion is very swift euen euery two houres about a degree, then in any other of the Planets : and for reddyer performance heereof, the Horarium Planetarium in the 29 Cap. mentioned, shall yeeld you speciall helpe, if you imagine the 10 houres of the Arch of the sextans to be 10 degrees, and cause the planetary Index to be numbred for this purpose into 12 houres from the Centre outwards.

Now when you will rectifie the true place of any Planet (admit of the ☾) at noone any day proposed (admit the 5. day of March 1598) to another Meridian (admit to the Meridian of Quinsay.) First you shall get the semidiurnall gate of the Planet by taking the motion of the day proposed (vz. 2 degr. 40 min. in ♋) out of the motion of the day following (being the 6 day of March, and the Moones place or motion then 15 degr. 1. min. in ♋) and take halfe the remaine (vz. 6 degr. 10 min.) which you shall seeke in the arche of the sextans (as though they were 6 houres 10 min.) and thereto set the Index, there let him stay. Then as in the last Cap is taught, seeke the houre distaunce of the Meridian of the other place (vz. of Quinsay) from yours, vz. from that of Anwerpe (for Stadius) but for my Vranicum from London, (admit the houre distaunce 8 houres and an halfe) and that distaunce seeke in the Index planted as before by the numbers reckned from the Centre outwards vnto 12, and marke what paralell

D   cutteth

The Vranicall

cutteth the Index there (vz. at the sayd 8½ houre thereof) and follow it to the arche of the Sextans, which shall there shew you (4 degrees 30 minutes) the motion of the Planet answerable to that houre distaunce of their Meridians, which if it be Westwards from your Meridian, you shall adde to, if Eastwards (as heere it is) then subtract out of the place of the Planet found (vz. out of 2 degrees 40 minutes of ♋) and so shall you haue (28 degrees 10 minutes in ♊) the true rectified place of the Planet to that Meridian proposed.

Note, that for ♄, ♃, ♂, ♀, ☿, and the ☉, whose semidurnall gates seldome exceede minutes, you may take 3 or 4 houres of the Sextans for one degree, and worke as before most exact.

Cap. 8.

¶ How to know the declinacion and right ascencion of any Planet, fixed Starre, or point of the heauens, and their degree of Culminacion.

Here are in the Mater, or Celestiall of this Astrolabe, to auoide confusion, only two sorts of circles delineated, & those are the circles of the-clipticks longitude and latitude in the 2. Cap. mentioned, issuing from his Pole D to his opposite. And the very like vnto those ought to haue been delineated, issuing from the Poles of the world, limitting the right assentions and declinations of the Starres, in regard of the Equinoctiall, and are commonly called Meridians and Paralells, which because the nature of this proiectement doth cast the Meridians into streight lines issuing from the Poles, both lodged in the Centre vnto euery of the 360 degrees of the Equinoctiall, or rather of the limbe and the parallells into concentricke circles, they are therefore supplyed by playing about the radiall degrees

Astrolabe.

of the noone-lines on the Centre, by the degrees of the limbe, and might as well be done by a thred & pearle applyed to the radyall degrees of the Scale planted at the beginning of ♋.

Further you shall vnderstand that countrey is sayd to haue a right Horizon, vnto which some one of the Meridians is Horizon, that is to say, where the Æquinoctiall passeth by the Zenith or point directly ouer head, called the Verticall point, and both the poles of the world euen with the Horizon, and neither of them eleuated, all other Horizons where either Pole is eleuated, are called oblique Horizons.

Now therefore when you are to get the right ascension of any Planet-Starre, or point of the heauens assigned, you are but to learne what Meridian and Paralell cutteth the same, and if you do but lay the noone-line thereon, it sheweth you in the limbe the right ascention, and in the scale of the Radiall degrees on the noone-line the declination. For example, the Sunnes Apex placed at the 26 degree 3 minut. of ♓ by the 2 cap. lay the noone-line thereto, and it shall shewe you in the limbe 356 degrees 25 minutes his right ascencion, and among the radiall degrees of the noone-line 1 degr. 30 min. his declinacion, and that Southwards of the Equinoctiall. For the 5 day of March 1598 likewise apply the noone-line vnto the chiefe Carect of ♄ there placed at 2 degr. 27 min. in ♎, and in latitude 2 degr. 44 min. North, and it shall shew you in the limbe 182 degrees his right ascencion, and among the radiall degrees, 1 degr. 10 min. his declinacion South from the equinoctiall the same 5 of March 1598. Againe, if for any Starre, as admit the great Starre of the first light in or neere the eye of the Bull, called Oculus tauri, the noone-line layd thereon, sheweth in the limbe 63 degrees 30 min. his right ascencion, and 16 degrees his declinacion North from the equinoctiall, though he be in latitude South from the-clipticke, and it cutteth in the Zodiacke the 5 degree 35 min. of ♊ his degree of culminacion: for alwayes y̆ degree of the-clipticke which coascendeth or cutteth

The Vranicall

any Meridian or right Horizon with any Starre, or other point of the heauens, is called the degree of his Culminacion, because the Meridian of any place is called Culmen Celi.

Cap. 9.

¶ How to finde the true Meridian in any countrey by helpe of this Astrolabe.

HE Meridian circle of any latitude, is that great circle of the Sphear which passeth by the Zenith or verticall point of that place, and a line drawne by his two intersections with the Horizon there, is called the Meridian line. Also, that Starre which in this Astrolabe you see in the outtermost tip of the tayle of Vrsa Minor, is commonly called the Pole starre, in Latin Stella polaris, or Cinosura, which in this age is some 3 ½ degree distant from the true North point or pole of the world, although the ordinary sort take it to be the Pole it selfe, yet in euery age to come by reason of the foresaid sliding backe of the equinoctiall points, he shall draw neerer and neerer to the Pole vntill the yeare 2220 after Christ, and then shall he be but 30 minutes from the Pole, and his longitude shall be then 0 in ♋, from whence he will then begin to elongate againe from the Pole till he come to 0 in ♑ about the yeare 15520, at which time his distance from the Pole shall be 47 degrees 30 minutes. And whereas Petrus Nonius sayth, that in times past he hath been 12 degrees 24 minutes from the Pole, and is now 3 degrees 30 minutes, happily he meaneth from the creation of the world.

This Pole-Starre in our age is found to culminate with the 6 degr. 20 min. of ♑, and to be 88 degr. 24 min. declining North from the equinoctiall: on which 6 degr. 20 min. of ♑ if you lay the Noone-line, you shall first thereby see what

Starres

A strolabe.

Starres do continually culminate or come vnto the Meridian with it, and you shall finde it to cut on the faire Starre of the third light, in the pummell of the chayer of Cassiopeia, and another of the same light the northermost of the girdle of Andromeda, and in the opposite part, vz. by the North line, you shall see the 60 degr. 20 min of ♎ to culminate almost with the innermost of the three in the tayle tayle of Vrsa maior called Alioth, and two of the Starres of ♍, But that of Cassiopeia, and that of Vrsa maior are best for this purpose. Hauing thus by your Astrolabe found out a Starre or two which haue the same degree of culminating any Meridian as the Pole-Starre hath, now shall you neede but in any starry night to hang vp some line, or erect some round pole or staffe perpendiculer to your Horizon, and then watch the going about of the Starres, vntill some one of those two Starres come with the sayd Pole-Starre very euen and directly with your perpendiculer, for then be sure that those two are both directly in the Meridian: Wherefore you shall then speedily plant another perpendiculer iust betweene your first and the sayd two Starres, and there let them remaine continually, for they shall serue you for many good purposes at any time after both day and night, in stead of a Meridian, I could set you downe many wayes more, but this is most pertinent to this Instrument, and is of all other the readyest, especially whē as by the 22 Cap. following you shall beforehand know what houre these Starres will come to the Meridian, that you neede not watch long for them.

Cap. 10.

¶ How to attaine the latitude or Poles eleuation in any
Countrey vnknowne, and to choose out the Verticall point in the noone-line, the Horizontall
point in the North line, with the Zenith
line seruing thereunto.

The Vranicall

He Latitude of any place, is the distaunce of his Verticall point from the Æquinoctiall, counted by the degrees of his Meridian circle on the South part, and is alwayes equall to the eleuacion of the Pole of the world aboue the Horizon counted in the same Meridian on the North part, either of which can neuer exceede 90 degrees: for which cause, there are in the Zenitfer 90 Zenith lines described, to serue euery turne. And the reddiest and most sure way to get the latitude in any strange place, hauing by the last Cap. setled your Meridian, is to take the Meridian altitudes of some one North Starre that neuer commeth South of the Zenith there, both aboue and vnder the Pole, then take the greater Altitude out of the lesser, and of the remaine adde the one halfe to the lesser, so haue you the latitude or Poles Eleuacion sought, whose complement is the Equinoctials eleuacion aboue the South point of the Horizon.

For example, admit the first Starre of the tayle of Vrsa maior called Alioth, if you obserue his comming to the Meridian aboue the Pole-Starre, and there finde his altitude to be 84 degrees high, and afterwards finding him againe in the Meridian vnder the Pole (which will be some 12 houres betweene) and there but $19\frac{1}{3}$ degrees high, then take $19\frac{1}{3}$ out of 84 there resteth $64\frac{2}{3}$, halfe that being $32\frac{1}{3}$ added to $19\frac{1}{3}$ maketh $51\frac{2}{3}$ our eleuacion heere at Reading, and so much is our Zenith, or Verticall point, all is one, distant from the Equinoctiall, and that $51\frac{2}{3}$ taken out of 90, leaueth the complement $38\frac{1}{3}$ the Equinoctials height aboue the Horizon in our Meridian.

Now, if you will choose out in the noone-line of the Zenitfer the Verticall point fitting thereunto, recken thereon from the Equinoctialls intersection, with the noone-line by the

radiall

Astrolabe.

radiall degrees and numbers thereon set betweene the Centre and Meridies, your latitude found, for there shall be your Verticall point desired (vz. for our Eleuacion heere at Reading, the Pole being 51 ¾ degrees high, I reckon that by the numbers in the noone-line, and there I conclude is our Verticall point, or of any Countrey hauing like latitude.) Likewise reckon on the North line from the Centre towardes Media nox, your sayde latitude founde (vz. the said 51 degrees 40 minutes,) and there shall be the place or point of your Horizons intersection, with the North part of your Meridian (51 degrees 40 minutes vnder the Pole) which I call the Horizontall point. And betweene the Verticall and Horizontall points thus founde, you shall alwaies finde 90 Radiall degrees complete and no more, which 90 degrees are the true intersections or crossings of the 90 Almicantares of that latitude with the North part of the Meridian, which for breuities sake I will heereafter call the North degrees, and those 90 degrees of the noone-line from the Verticall point towards Meridies, I will call the South degrees, both which in a particular would be numbred from the two Horizontall points, vnto the Zenith or Verticall point, there ending at 90 accordingly.

Now for the Zenith line in the Zenitfer appertaining to your latitude found, it is easily chosen by the numbers ascribed to the inner ends of those 90 Zenith lines, as in the 1 cap. is shewed, which 90 inner ends are y 90 Zenith points, and do answere by the fiduciall line of the Cursor, vnto the 90 Verticall points of the noone-line, (vz. the 51 ¾ Zenith line must be chosen for our latitude heere at Reading,) euery of them carrying his 90 Almicantare rootes distinguished by the curued lines, and numbred from their outermost end inwards towards their Zenith points, as in the 1.Cap. is shewed.

Your Verticall point, Horizontall point, and your Zenith line appertaining thus found, you must in any wise marke them specially from the rest except your memory be good, the reddier to finde them in all your actions.   Cap.

The Vranicall

Cap. 11.

¶ How to take the Meridian Altitude of the Sunne, or any Planet or Starre, and thereby to get the latitude in any Countrey.

It is a matter very easy and common, either by a Quadrant with a line and plummet, or by the circle or this Astrolabe hanged perpendiculer, furnished with a rule and two sightes to take an altitude: for the Sunne, by suffering the Sunne-beames to pearce the sight-holes, and for Starres by applying the sightes to your eye, which altitude of the Sunne or any Starre being taken in the direct South or North by help of your Meridian set by the 9 cap. is called the Meridian altitude. But if you haue no Meridian set, place your Quadrant or Astrolabe towards the South or North, and thereby obserue the seuerall altitudes of the Sunne or Starres as they rise higher and higher, and the highest shall be the Meridian altitude. Where note, that such Starres whose declinations exceede the latitudes complement, do come twise vnto the Meridian at 12 houres betweene, yet neuer come South of the Zenith, or Verticall point.

The Meridian altitude thus had (As for example heere at Reading the 2 of October 1596 about 3 of the clocke in the morning, I obserued the comming of the Starre oculus Tauri to the Meridian, and there tooke his Meridian Altitude by a Quadrant 54⅓ degrees) take it out of 90 degrees (vz. 54⅓ out of 90) so haue you (35⅔) the complement or verticall distance of the Planet or Starre (oculus ♉) proposed, from the Zenith of your being. Now if this Meridian Altitude happen on the North from your Zenith, then must you place

the

Astrolabe.

the North line of your Zenitfer on the centre of your Planet or Starre found out in this Astrolabe: but if it happen South of your Zenith, (As heere this Meridian altitude of ocul. ♉ did from our Zenith at Reading) then bring the noone-line thereto. And from that touching of the Starres centre therewith, (which you shall finde to touch at 16 degrees, as they are on the noone-line there numbered) you shall recken inwards towards the centre of the Celestiall so many degrees as the sayd verticall distaunce amounteth unto, (vz. the saide 35 ⅔ degrees) and the end of that reckoning shall assuredly shewe you (51 ⅔) in the noone-line the Zenith point of your being, whose number ascribed (vz. the saide 51 ⅔) is your latitude desired, (thereby I might conclude, that the latitude heere at Reading was 51 ⅔ degrees) whose Horizontall point and Zenith line you may easily finde out by the last Cap.

Cap. 12.

¶ How to know what day of the yeare it were if it were forgotten.

Et the Latitude by the 10 or 11 cap. (admit our latitude 51 ⅔ degrees) and the meridian altitude of the ☉ by the last cap. (admit 25 degrees) recken the complement thereof (vz. 65 degrees) in the noone-line from the Verticall point towards Meridies, and looke what radiall degrees it falleth at (you shall finde it at 13 ½ degrees Southwards from the Equinoctialls intersection) and apply that degree by moouing about the Zenitfer unto the elipticke circle, according as you gesse the time of the yeare to be, vz. If you thinke it to be betweene Midsomer and Christmas, then apply the same 13 ½ degr.) unto the autumnall hemicicle, where ♋ ♌ ♍ ♎ ♏ ♐ are, (where it must needes touch upon the 24 degree of ♍) if betweene Christmas

The Vranicall

mas and Midsomer then vnto the vernall hemicircle of the others Signes (where it must needes cut the 6 of ♏) and there shall most assuredly be the true place of the Sunne for that day. Which being had, then you may easily by the 5. Cap. euersed as well finde the day answerable thereto, as there you found the Sunnes place answerable to the day vice versa. Prouided that the Sunnes place be rectified as in the 6. Cap. if needs be.

Cap. 13.

¶ How to know what it is a clocke at any time either night or day, by the Sunne, Planets, or Starres.

Ou must alwayes intend that the Planets are weekely placed, as in the 3. Cap. is taught. Also obserue this rule generall, to keepe the nooue-line of the Zenitfer still vpwards, by turning about your Astrolabe, imagining your face turned still to the South, except when you are specially to deale with the North line, then keepe him vpwards, and then imagine your face turned towards the North. Now when you desire to knowe the houre of the day or night, you shall as in the last Cap. is mentioned, either by a Quadrant, or this Astrolabe, take the Altitude of the Sunne at any time in the day or in the night of any Starre. (For Example this yeare 1596 the 2 of October in the morning at my poore house at Swallowfeyld by Reading, to see how the time passed, I tooke the altitude of the Sunne 15 degrees high, And the same euening I tooke the altitude of the great Starre Arcturus 10 degrees high towards the West) Which had, then shall you prepare your Almicantifer vnto that altitude, thus: In the Zenith line chosen by the 7. Cap. for your place (vz. in the 51⅓ Zenith line for vs heere about Reading) seeke out among his Almicantare rootes the number of that Altitude taken,

Astrolabe,

taken, (vz. in the morning I sought 15 for the ☉, and at night 10 for Arcturus) & there plant the Cursor, then reckon on the North line frō your Horizontall point found by the 7 inwards towards the centre the same altitude againe (vz. frō the Horizontall point of our latitude, being the 51 ½ deg. of the Northline counted from the center, I reckoned inwards for the ☉ 15 degrees, and for Arcturus 10 degrees, the one fell out at 36 ⅔ degrees from the center, the other at 41 ⅔) and thereto, set the pointer of the Almicantifer, which pointer and Cursor so set, I call the Almicantifer prepared: Then leade about the Almicantifer, which must also leade after him the nooneline, vntill the pointer come to touch exactly either the Apex of the Sunnes Carect, or the very centre of the Starre by which soeuer you deale, and this I call the applying of the pointer to any Planet or Starre: There hold the Zenitser fast, and slide downe the Cursor to the centre, applying there his Fiduciall line labell-wise to the Sunne, which shall shew you in the houre circle the houre desired, and this I call labelling of the Sunne to the houre circle: Yet vpon these conditions, that if you deale for the East Hemispheare, that is to say, if your altitude taken were on the East part of the Meridian, vz. betweene the North and South Eastwards (as for the ☉ it was) then in this leading about shall the Almicantifer play leftwards of the noone-line towards 6 of clocke in the morning of the houre circle which I call the East Hemispheare, and so to leade the noone-line after it leftwards, but if your altitude taken were on y̆ West part (as for Arcturus it was) then shall the Almicantifer play and keepe on the right side of the noone-line towards 6 of clocke at night, and leade him about rightwards after him (According to which precept, the said 2 of October in the morning, I applied the Almicantifer his pointer prepared as is said vnto 15 deg. to the Apex of the ☉ (placed as is said at 20 ⅔ of ♎) in the East Hemispheere, and there did I labell to the Sunne, which shewed me halfe an houre past 8 in the morning, the houre desired.

Likewise

The Vranicall

Likewise at night I applied the pointer prepared, and planted for 10 degrees vnto Arcturus in the West Hemispheare, because Arcturus was in the West, and then againe I labelled to the ☉ which shewed me in the Houre circle halfe an houre past 7 of the clocke at night, the houre desired: For alwayes (Locus Solis ostendit horam.) And I dare boldly say, that neuer was seene before this more facillitie to know the houre by the Starres, by any other Astrolabe, or Instrument what soeuer.

Cap. 14.

¶ Another more easy way to finde the houre by the Starres.

Auing your Meridian planted about your dwelling place, as in the 6 Cap. is taught, see into the element if any Starre you know well (as by the 25 Cap. you shall easily know them all) be euen therewith either on the South or North part. And vpon that Starre sought in the Celestiall, place accordingly the North line or noone line (either of them now esteemed from the Zenith point) and there labell to the ☉ so shall the houre sought shew it selfe in the houre circle vnder the fiducial line of the Labell or Almicantifer now vsed Labell-wise after the woonted manner, this needeth no example.

If you haue no Meridian perpendiculers set, you may at any time presently set them with two stickes by the Pole-Starre it selfe sufficient for this purpose. I my selfe alwayes kept a standing marke ready set in such sort, that some tree, chimney top, or gable end, should be directly South from the same. If no Starre be in the South or North, it is not hard to gesse the distaunce of the neerest you see, and place your noone or North line accordingly, especially by two Starres, on either side one, to marke what part of their distance your

Meridian

Astrolabe.

Meridian cutteth, and accordingly to place the noone or North line betweene them.

Cap. 15.

¶ Another easy way by the ☾.

Ake the houre which the shade of the ☾ giueth on any Sunne-dyall, seeke the same houre in the houre circle, and labell it to the Moones Carect planted by the 3. Cap. and then holding fast the Zenitser, labell to the Sunnes Carect, which shall shew in the houre circle the houre desired.

Cap. 16.

¶ How to know the meridian Altitude of the Sunne, or any other Planet, Starre, or point of the heauens in any countrey, and what houre of the day or night any of them shall culminate, or come to the Meridian of that place.

HE Carects being planted by the 3 Cap. you shall but applye the noone-line to the Apex or Centre of the Planet, Starre, or point desired, and the degrees thereon found betweene the Uerticall point and it taken out of 90, leaueth the Meridian altitude desired, whether it be North or South, and there also if you labell to the Sunne, it shall shew you in the houre circle the time when that Planet, Starre, or point, commeth to the Meridian. (For example, the noone-line brought to the Sunnes chiefe Apex, being planted by the 3 cap. at 26 3.min. in ♓, his place for the 5 of March 1598. you shall finde him

E 3   by

The Vranicall

by the degrees of the noone-line 53 degrees distaunt from our Verticall point heere at Reading therein, vz. from the 51⅔ degr. thereof, which 53 taken out of 90, leaueth 37 the Meridian altitude. In like manner for any Starre, Admit oculus tauri the noone-line applyed to his middle pricke, you shall finde it distaunt from our Verticall point 35⅔ degr. that taken out of 90, leaueth 54⅓ his Meridian altitude, As if you make tryall by obseruacion, as in the 8 Cap. is taught, you shall finde it so. And if you there labell to the ☉ you shall finde it to cut in the houre-circle 4 of clocke in the after noone, and almost halfe an houre past, which is the time that oculus tauri shall come vnto the Meridian the sayd 6 of March 1598. But when you worke for the Sunne as before, he onely is himselfe the pointer of houres, and therefore must needes be in the Meridian at Noone, or 12 of clocke. But if it be for any North Starre that neuer setteth, and that you would know his meridian altitude vnder the Pole, apply the North line thereto, and the degrees thereof betweene it and the Horizontall point is your desire, and there also labell to the Sunne, and it shall shew you the houre.

Cap. 17.

¶ How to know the houre of Sunne-rising and setting, and the length of the day and night, together with his oblique ascencion, and his difference of ascencion any day in the yeare.

Our day and yeare proposed (Admit the said 8 day of Iune this yeare 1596 in the 3 Cap. mentioned, when our most noble Lords lauched from Plimmouth towardes Spayne) and the Sunnes Apex placed by the 3 cap. (vnto 27 degrees 58 min. of ♊) Then shall you set the Cursor vnto the outer end of the Zenith line

A strolabe.

line of your countrey, (vz. vnto the .o. almicantare roote of the 51 Zenith line being for the latitude of Plymouth,) and there fastening the Cursor, apply the pointer vnto the Horizontall point (of Plimouth, being the 51 degree of the North line) and this I call preparing of the Zenither to the Horizon. Now, first to begin for the Sunne-rising, being thus prepared, you shall according to the precept in the 9 Cap. let the Almicantiser play on the East Hemicicle, and there apply his pointer to the Sunnes Apex (at 27 degr. 58 min. in ♊) there hold fast the Zenitser and Labell to the Sunne, so shall appeare in the houre circle (4 of clock in the morning, and somwhat lesse) the time of the Sunne-rising for your (8 of Iune 1596 at Plimouth) day and yeare proposed. And if there also you labell to the East point or 6 of clock in the morning of the houre circle, it shall shew you in the limbe of the Celestiall (56 degr. 20 min.) the oblique ascencion of the Sunne for that time and place he is in, which taken out of the right ascencion of his place (of 27 degr. 58 min. of ♊, being $87\frac{1}{2}$ degr.) had by the 8 Cap. there shall remaine (31 deg. 10 min.) the difference of ascencion. In the very like manner worke for the Sunne-setting, by letting the Almicantiser to play on the West Hemicicle, and there apply the pointer againe to the Sunnes Apex (vz. to the foresaid 27 degr. 58 min. of ♊) and there labell to the ☉, so shall you see in the houre circle (some 10 min after 8 of clocke at night) the houre of Sunne-set: and the space of houres numbered betweene Sunne-rising and setting (being 16 houres $\frac{1}{5}$) is the length of the day, and those ($16\frac{1}{5}$ houres) take out of 24 houres, leaueth ($7\frac{4}{5}$ houres) the length of the night. Also the houre of Sunne set (vz. $8\frac{1}{10}$ doubled any day, maketh ($16\frac{1}{5}$ houres) the iust length of the day as before. And heere also if you labell to the West point of the houre circle, it shall shew you in the Celestialls limbe (118 deg. 30 mi.) the oblique descencion (in the said 51 latitude) As for the difference of descencion, it is all one with the difference of ascencion.

This

The Vranicall

This Chapter is most easily to be performed by the particular moouer in the 1 Cap. mentioned: For if you mooue it about leftwards (vz. from West towards East, according to Copernicus Hipothesis, and contrary to all other Astrolabes) till the East part of the particular Horizon come to the Sunne, the Labell layd thereon, sheweth the houre of rising, and continuing that moouing till the Wester part of the Horizon come to the Sunne, the Labell there applyed, sheweth the Sunne setting: yea, and more then that, the Labell there at both places sheweth the amplitude of the Sunne rising and setting, by the prospectiue degrees of the Horizon. Note, that if you would be curious, you should at Sunne rising apply the Labell a quarter of a degree short, and for the Sunne setting a quarter of a degree further on in the clipticke, then the Sunnes Apex is planted in regard of the Sunnes diurnall motion, being euery 24 houres almost a degree, for that the Apex is set but iust for noone, the reason whereof is shewed in the 6. Cap. Note also, that for the rising and setting of Starres, and the other Planets, they are to be done by the particular in all respects as for the Sunne, sauing that you must remember semper locus solis ostendit horam: Also, the getting of the assencions and descencions haue all one working with the generall.

Cap. 18.

¶ To know the houre of rising or setting of any other
Starre, Planet, or point in the Heauens, and their
diurnall and nocturnall arches, and their
oblique assencions and difference
of assencion in any
Countrey.

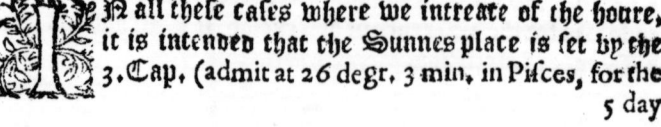
IN all these cases where we intreate of the houre, it is intended that the Sunnes place is set by the 3. Cap. (admit at 26 degr. 3 min. in Pisces, for the

5 day

Astrolabe.

5 day of March 1598.) Now to knowe the time of rising or setting of any Planet, Starre, or point proposed (admit the great Starre Arcturus) for any time and place proposed, you shall prepare the Almicantiſer to the Horizon proposed in all respects as in the last Cap. you did (vz. the Curſor planted vnto the outermost end of the 51 ⅔ Zenith line, and the pointer vnto the 51 ⅔ Horizontall point ſeruing with vs heere at Reading.) And then for the time of rising, you must let the Almicantiſer play on the East Hemicicle, and there apply the pointer to the centre of (Arcturus) the Starre or point proposed, and there holding fast the Zenitſer, labell to the Sunne, ſo ſhall you finde out in the houre circle (about 26 min. past 5 of clocke in the afternoone) the houre and time of his rising. And there also if you labell to the East point of the houre-circle, it ſheweth in the limbe of the Celestiall (280 ¼ degrees) the oblique aſcencion of the same Starre or point (Arcturus) in the (51 ⅔) latitude proposed, which taken out of the right aſcencion (209 ¼) had by the 8 Cap. leaueth (19 degrees) the difference. Note, that if happily you thinke your houre circle with the ſmalleſt to direct you to the oblique aſcencion exactly when you lay the Labell on the East or West points, know that he must cut the like part of the Degree where he lighteth as the noone-line or North-line do in the degrees where they are. Note also, that to knowe the oblique aſcencion, is as much to ſay, as to know what degree of the Æquinoctiall doth riſe with any Starre or Planet on any oblique Horizon. As for the houre of ſetting, let the Almicantiſer prepared as before, play on the West Hemicicle, and aply the pointer there againe on the ſaid Starre (Arcturus) or point proposed, where if you labell to the Sunne (at 26 degrees 3 min. in ♓,) it ſhall ſhew you in the houre-circle (20 min. past 10 the next morning) the time of his ſetting: and the houres reckoned on the houre-circle betweene his time of riſing and ſetting (being 16) is his Diurnall arch, which taken out of 24, leaueth (8 houres) his nocturnall arch.

F   Cap.

The Vranicall

Cap. 19.

¶ How to set perticular degrees on the Almicantifer, for any perticular latitude.

Do not stand making speciall Chapters of the vse of the perticular mouer in the 1 Cap. mentioned, because it consisteth but of the very same Horizon circle and 12 houses of one perticular latitude, which old I. Stophlerus, and our old english Laureat G. Chaucer, haue so largely and plainely written of: withall, being a thing most commonly in vse, and differing in nothing, but that they according to the auncient Astronomers, appointed the Starry Heauens to moue rightwards from East towards West, vppon the earth or fixed Horizon of the place. And I according to Copernicus cause the earth or Horizon to moue leftwards from West towards East, vppon the Starry Firmament fixed: In so much, that if in this my Astrolabe you hold still that perticular moouer with one hand, and with your vnder hand turne about the Celestiall, then is it iumpe Stophler againe. In which motion (a pretty thing to note) one that standeth by shall hardly perceiue any other but that the Reete mooueth, although in deede you turne about the Mater, strongly confirming Copernicus Argument, who sayth, that the weakenesse of our senses do imagine the Heauens to moue about euery 24 houres from East to West by a Primum mobile, where as in deede they haue been alwayes fixed, and it is the earth that whirleth about euery 24 houres from West to East, of his owne propper nature allotted vnto him, as most fit for the receptacle of all transitory things, being appointed in a place where nothing is to stay him from his continuall moouing: but of those things I haue more iust occasion to intreate of in my Organum V-
ranicum

Astrolabe,

ranicum very shortly to come foorth. But now if the sayd particular moouer should also haue his perticular Almicantares like the Reete of my Iewell tediously cut out, they would much obscure this faire face of the Celestiall, very needlesse: because the perticular Zenith-line of the place, with his 90 rootes annexed to the noone-line for the Cursor to play vppon, and the perticular degrees by this Cap. prepared on the Almicantifer, shall serue in all respects as well, or rather better. And the manner of preparing them is in a manner already set downe in the 13 Cap. Yet is it more easily to be done by a perticular, whose radyall degrees on the North line, which before I haue tearmed North degrees, are or should be ready numbered from the Horizontall point, towards the Verticall point, ending there at 90 degrees, which in the generall can not be suffered. Wherefore keeping the fiduciall line of the Almicantifer euen with the North line, then sliding the Cursor from roote to roote of the 90 rootes of your perticular Zenith line, set nickes in the Almicantifer at euery degree, from degree to degree of the 90 North degrees, numbered from the Horizontall point, towards the Verticall point, correspondent to each roote as they passe in order, till the whole 90 particular degrees be in manner as in the 13 Cap. is shewed by those nicks placed on the Almicantifer, and then distinguish them by tennes, and set numbers vnto them accordingly, so are they very ready for diuers speciall purposes in one latitude, and if neede were, a man may by the two edges on both faces set the like in for foure latitudes, and those will neere be sufficient for any one region or prouince, or else he might haue it made with some chanell to receiue a slip of velome or paste-boorde, whereon with pen and inke speedily to prepare it for any latitude.

The Vranicall

Cap. 20.

¶ How to know the amplitude of the rising, or setting of the Sunne or Starres.

HE true East and West points of any Horizon are the Intersections of the Equinoctiall therewith: and that is the cause why in describing the generall Horizons, they make them all to meete at two points which are those, and for that cause it is manifest, that there can no Planet, Starre, or point of the heauens, rise or set directly in the true East or West points of any Horizon, but only such as are scituate for the time being in the Æquinoctiall circle. Wherefore those that are North from the Equinoctiall, must needes rise and set Northwards from the true East and West points, and those that are Southwards rise and set Southwards, and that more or lesse, according to their distaunce, and the obliquitie of the Horizon proposed. And the degrees of the Horizon circle betweene the true East and West points, and their rising and setting, are called the amplitude of rising and setting, which in the perticular moouer are immediately seene by the degrees of the Horizon circle, as in the 17 Cap is sayd: but by the generall it asketh a little paines in this manner. Get the oblique ascencion of the Sunne, Starre, or point proposed for the time and place, by the 17 or 18 Cap. (which for the Sunne was in the Horizon of Plymouth the day of the prosperous setting foorth of our saide noble Lords, founde in the 17 Cap. 56 degrees 20 minutes of the Celestiall limbe,) and thereto apply the noone-line, and there hold the Zenither fixed: where if you haue the Almicantifer by the last Cap prepared for the first latitude, vz. vnto the Zenith line, neerest vnto the noone-line,

whose

Astrolabe.

whose Horizontall point is in the centre of the Astrolabe, and then runing the Cursor downe by the rootes of the same first Zenith line, keeping the fiduciall line of the Almicantifer still on the Sunne, Starre, or point proposed, vntill the Cursor among the rootes, and the ☉ or Starre among the prepared degrees doe cut like number, for then is the complement thereof, the degrees of the amplitude sought. But if your Almicantifer be not prepared, then must you prepare him and his pointer as in the 13 Cap. is taught, first to one degree of altitude, then to another, making still proffers to the ☉ or Starre, till the pointer touch vppon it, (which will fall out at 41 degrees of that first Zenith line and latitude) and the degrees so found, taken out of 90, leaueth (49 degrees) the amplitude (of the ☉) sought.

Cap. 21.

¶ To know what Starres do neuer rise or set in any latitude, and which do once a day touch the Zenith point, or the Horizontall point.

His matter is most easy, (your Verticall point and Horizontall point chosen by the 10 Cap.) to be done by the bare moouing about of the Zenither, for in that motion, such Starres (as the tayle of Vrsa maior, and the head of Perseus, and such like) as touch the Verticall point be sure (in our 51 ½ latit. heere at Reading) do come once in 24 houres to the Zenith of that place: And such as do not reach vnto the (51 ½ Horizontall point, (as all Vrsa maior but 2 Starres, Vrsa minor, Cassiopeia, Cepheus &c) do neuer set. And such as touch vppon the Horizontall point, (as Caput Medusæ lucida lyre, the ring of Andromedaes cheine &c,) do once in 24 houres touch vppon the Horizon of

The Vranicall

the (51⅔) latitude proposed, yet do not set vnder it. By the particular moouer, these things are most plainely perceiued, and reddily done by the moouing of the Horizon circle about.

Cap. 22.

¶ To know any day and houre what Starres or points in the Heauens are in or neere the Meridian, or do rise or set in any Latitude.

His matter is as easy as the last. Labell the Sunne (admit placed at 20 degrees 10 min. in ♎ for the 2. of October 1596) to any houre in the houre circle (admit to 8 of clock at night) and so shall the noone-line of the Zenitfer shew (the two Starres, called Caude Capricorni, and Lucida Hidri, but a little past, &c.) such Starres as are the same instant in the South, and the 12 degree of Aquarius in the medium Celi vnder the noone-line. And there holding fast the Zenitfer, being prepared vnto the Horizon (51⅔ heere at Reading) proposed, leaue the Almicantifer about the East Hemispheare, and looke what Starres (as part of Perseus, one of the heads of ♊, the Bulles head, the finne of the Whale, &c,) you shall finde the pointer to touch vppon: the same assuredly, do euen that moment emerge and rise on the Horizon. That done, the Zenitfer still remayning and prepared to the Horison as before, leaue the Almicantifer about the West Hemispheare, and you shall finde the pointer to touch vppon such Starres (vz. one in the arme pit, another in the bowe of ♐, &c.) as are euen then ready to set and go downe vnder the Horizon. This is also most easy by the Horizon circle of the particular, which being but set to the houre, the Starres rising appeare in the East Hemispheare, and those setting: in the West and those in the Meridian: vnder the noone-line.

Cap.

Astrolabe.

Cap. 23.

¶ How to know the houre distance of any Planet, Starre, or point of the Heauens from the Meridian, any day and yeare proposed, and in any latitude.

That is to say, how many houres any Planet, Starre, or point (admit oculus Tauri) not being in the South at any time assigned, (admit at 10 of clocke at night the 5 of March 1598.) will be before he come to the Meridian of any place (admit our Meridian heere at Reading in the 51 ½ latitude) proposed, or if he be past, how many houres he is past: You shall labell the houre giuen in the houre-circle (vz. 10 of clocke at night) to the ☉ planted as in the 3 Cap. (at 26 degr. 3 min. of ♓) there holde fast the Zenitser, and labell to your Starre (oculus Tauri) or point proposed, so doth the labell shew in the houre-circle (fiue houres ⅔ almost Westwards, and therefore past the Meridian) the houre and part that the Starre or point is distant, which if it be on the East part, then is not that Starre or point come to the Meridian, if Westwards (as heere it was) then is he past the Meridian (5 ⅔ houres) so many houres.

Cap. 24.

¶ How to know the height of the Sunne, or of any other Planet or Starre, for any houre of the day any yeare in any latitude.

Seeke your houre proposed (admit 10 of clock in the morning the 2 of Octob. 1596 in our 51 ½ lati.) in the houre-circle, & labell it to ye Suns carect (at 20 d. 10 mi. in ♎)

there

The Vranicall

there hold the Zenitfer fast. Now if the particular degrees be planted on the Almicantifer, as in the 19 Cap. is taught, you shall, (keeping the fiduciall line of the Almicantifer on the Starre or Sonnes Aper,) neede but to slide the Cursor too and fro on the Zenith line of the (51 ¾) latitude proposed, vntill the Cursor among the rootes of that (51 ¾) Zenith-line, and the Sunnes Aper or centre of the Starre (admit of oculus ♉) among the particular degrees of the Almicantifer, do cut like number, (which you shall finde for the ☉ to be at the 25 roote, and 25 particular degrees, and for oculus tauri at the 3 ½ roote, and 3 ½ particular degrees,) for those shall be the degrees of the Sunne or Starres altitude sought. But if your Almicantifer haue not the particular degrees, then must you as in the 13 Cap. is taught, prepare it to some degree at a venture, and make a proffer to the Sunne or Starre, to see if the pointer will touch it: if it do not, prepare it againe and againe if neede be, further off, or neerer, as the cause requireth, till it do hit, as in the 20 Cap. you did, which with a very little practise will be as easy as the particular. This and the 20 are the onely two harvest in the booke, and therefore spare not so little paines as they require.

Cap. 25.

¶ How you shall most easily learne to know all the Constellations and Starres in the element it selfe.

Euen as a lofe of dowe or paste, after it is new moulden round and copped, if you then put it into a presse, will become a flat cake: Euen such a cake made of the round Sphaere or Globe pressed, is the Celestiall of our new Astrolabe, which I would haue you now for this purpose imagine to be a lofe or round Globe againe: and that every of his circles had

Astrolabe.

had their due conuexities in such sort, that a little pretty fellowe like Tom Thumbe might easily seate himselfe vnder them as vnder a Canapy in the very centre of the concauity, with this conceite, if at any houre of any night giuen (admit at 10 of clocke at night the saide 5 of March 1598) you hauing first placed that houre of the houre-circle vnto the Sunnes Carect placed by the 3 Cap. (vnto 26 degr. 3 min. of ♓,) there fastning the Zenithser: And then going abroade, by help of your Meridian set, as in the 9 Cap. do place your Astrolabe flat on some stoole, plaine leuell with the Horizon, with his noone-line directed towards the South, euen with the saide Meridian line, and the North line towardes the North. Then with a priuy sconce and candle, one while viewing your Astrolabe, another while hiding the light, and viewing the element, you shall soone perceiue all the Starres of any constellacion in the heauens which are then aboue your Horizon, in the like proportion one from another, and differing in their magnitudes, as in the Celestiall they are described: in so much, that your little pretty fellow sitting in the centre vnder this conceited canapy of Starres and images, if it were transparent as glasse or crystall, should there plainely see the Starres in the element, each answering his match in this canapy, by the visuall lines from his eye, especially if the noone-line be eleuated according to the Equinoctialls height, or latitudes complement, because the proiectment is polare.

But if this do not satisfye you sufficiently, then is best for you hauing placed your Zenitser as before by the houre proposed, to see by the 22 Cap. what Starres are vnder or neere the noone-line, and of which magnitude they are, (you shall finde at 10 of clocke the 5 of March 1598 the great Starre of the first magnit. in the brest of the Lyon, called Cor Leonis, directly vnder it, and the bright Starre of Hydra of the 2 magnitude not much past) and by the 16 Cap. learne their meridian altitude (which for Cor Leonis, you shall by the

16 Cap.

The Uranicall

16 Cap. finde 52 degrees, and for lucida hidri 33 degrees,) which had, go foorth with an Astrolabe or Quadrant with his sightes set to that altitude directly towards the South, and you shall wellneere finde your Starres desired euen with your sight holes, and you be not too long about it. Also at any time you may lay the noone-line to any Starre (admit Arcturus) you desire to know, and then labell to any houre of the night proposed, (admit 10 of clocke) then marke what degree of the clipticke the labell there cutteth, (you shall finde 1 degr. 30 min. in ♊) and seeke by the 5 Cap. what moneth and day thereof that the Sunne commeth to that degree, (you shall finde on the 12 day of May anno 1598) for on that day be sure to finde that Starre in the South at 10 of clock at night, where you may make triall of his Meridian altitude as before. And hauing thus or any other way attained the perfect knowledge of some fewe, you shall easily by them descrye the rest by their scituation vnto them, and imagined distaunces to others, some being in streight lines from two other, some making a triangular, some a quadrangular forme of sundry shapes: And this shall suffice an ingenious conceit in a fewe nightes to learne to know most of the constellacions and their Starres, who shall the whiles finde the time spent in heauenly pleasure and delightes.

Cap. 26.

¶ Of the 4 Cardines, and the sundry rising and setting of the fixed Starres with the Planets.

THE Meridian and Horizon of euery place do alwayes crosse each other in halfe on the North and South points of the same Horizon: which crossing, maketh of those two circles 4 semicircles, which are called the 4 Cardines. Vnto either of which when any Planet

Astrolabe.

Planet commeth, especially if he be accompanyed or strongly aspected with some other forcible Planet, Starre, or constellacion, he worketh speciall effects, whether it be in natiuities or other causes, especially at any such time, looke assuredly for some present alteracion of the weather in that Horizon, because there they do strongly affect and aspect the ascendent Horizon and Climat, either by coniunction, opposition, or quartile, which are the strongest: the other, as the △ and ✶, are connted of lesse force. And whosoeuer will take vpon him to iudge of Aspects or Eclipses, if he haue not an eye whether they aspect the Climat, or to iudge of the weather, and not specially regard the effects of the fixed Starres with the Planets, shall finde it to prooue but lost labour, and happily be counted a common lyar, as some of our vnskilfull Prognosticators are, and the rest little better thought of, because of their negligence or ignorance. But to come to my purpose againe, you shall vnderstand, that when any Starre riseth together with any Planet on the Horizon (if with the ☉, then is it called the Cosmicall rising and if setting at Sunne rising the Cosmicall setting, if rising the Sunne setting, the Acronicall rising, if setting with the ☉, the Acronicall setting.) The Cosmicall rising is easily knowne by applying the Almicantifer prepared to the Horizon vnto any Planet in the East Hemicicle: there hold fast the Zenifer, and play about the Almicantifer, and looke what Starres the pointer toucheth on in the East Hemicicle, be sure those do rise at the same time Cosmically with that Planet, and those he toucheth on in the West Hemicicle, do then Cosmically set with the same Planet, then rising. Likewise, if you put ouer the Almicantifer so prepared, causing the Planet to lye in the East Hemicicle, applying the pointer thereto, and there holding fast the Zenifer, do leade the Almicantifer about, then be sure that such Starres as the pointer lighteth on in the East Hemicicle, shall rise acronically with that Planet setting, and those in the West Hemicicle,

The Vranicall

shall set acronically together with the Planet setting: and what time of the yeare any of these shall happen, is easily conceiued by that hath been said in the last Cap. and elsewhere: For vnto the Starre proposed, if you apply the pointer prepared to the Horizon, & there holding the Zenithfer, do remooue the Almicantifer till the pointer touch or cut the Clipticke either in the East or West Hemicicle. Then by the Ephemer. or my Vranicum you may easily know when & what day any Planet shal come vnto those points of the-cliptick so cut, and the degree of the ☉ for that day also labelled, shewetl there in the houre-circle the houre of the day (the Zenitfer being neuer stirred all the while) when any such rising or setting shall happen.

For example, this yeare 1596 some 5 or 6 dayes together before the end of September, after long sitting all day writing this present booke, I was in the euenings hindred of my recreating walks by sodaine rising of the winde with showres of raine, though the dayes were reasonable faire, I therefore sought this my new Astrolabe, and found the Pleiades or 7 Starres to rise in the euenings cosmically with Iupiter, being retrograde all the while, ☿ also being then retrograde, for retrogradacion causeth rayne, especially in ☿ without all fayle more or lesse as he shall be affected or aspected. Also the 2 of October 1596 I was sodeinly ouertaken riding from Reading to my poore house at Swallowfeild at euen, with a most stormy rayne that continued an houre or two after very extreame, it made me looke my Astrolabe againe, where I found spica virguis the Sunne ♂ and ☿ had set all together acronically within one halfe houre or lesse that this storme began, and then an houre after that Iupiter was to rise as before cosmically with the Pleyades. Now I leaue it to our sufficient Prognosticators to iudge whether these might be efficient causes sufficient or no, and do commend this my Astrolabe vnto them as a most needfull thing for them, wishing them to haue a speciall regard of the rising, setting, and culminating

Astrolabe.

of the fixed Starres. The common countreyman seeing any day inclined to moysture, can say, that it will surely rayne when the Sunne and the winde come together, neither will it indeede rayne till then for the most part, how likely so euer. And wherefore then should not we beleeue the learned, that affirme the rising of Orion, Arcturus, Corona, &c. to prouoke tempestious weather, the Kid and Goate windes, the Hiades, Pleiades and Succule rayne, and especially happening with any forcible Planet or strong aspect, the more violent. Were it not for consuming the time, I could recite examples of all these and diuers other that I haue vpon like occasions noted to my selfe, but dare presume no further, least I should be billed amongst the common liers, as others are happily causelesse: for no doubt the Almighty hath the raigne in his hand, to alter all at his pleasure when they haue all done their best. Yet that which is most to be condemned in them is, that they agree seldome in one coniecture, which must needes bewray shamefull ignorance in some of them at the least.

Cap. 27.

¶ Of the 12 houses, and setting the Figure.

Earned men considering the Sunne not only at his comming vnto the 4 chiefe points of the elipticke, vz. the two Equinoctiall and the two Solsticiall points, to cause the Spring, Autumne, Sommer and Winter, but also that in euery other of the 12 Signes he maketh some alteration of the yeare and seasons. Euen so, finding such speciall effects to happen by the comming of the Planets vnto the 4 Cardines, either with the fixed Starres or without, as in the last Cap. is touched, they did imagine that the Planets and Starres might haue also some forces and effects in other meane places betweene: and therefore tooke vpon them to adde 8 semicircles more vnto the 4 Cardines,

The Uranicall

Cardines, crossing with them at the two meridian Horizontall points, making vp the number of 12, which they call the 12 houses. Those on the particular moouer are ready drawne, so that applying but the houre in y moouers limbe to y Sunne, your Figure is set immediately and most excellently by this Astrolabe, fully fraught with all his needfull tackling, his Lords, termes, faces, triplicities, aspects, Planets and Starres in their places, natures of the 12 houses, and of the 12 Signes, the gradus masculini, feminini, lucidi, tenebrosi, vacui, fumosi, &c. and what not? appertaining to the Iudiciary arte, as if you reade Alkabitius, Claud. Dariotus which is englished, Gwydo Bonatus, and others their introductions to that arte, comparing the 2 Cap. therewithall, you will easily conceiue.

But to finde these 12 houses in any countrey (admit heere at Reading) by the generall for any day and yeare proposed, (admit at 10 of clocke the 5 day of March 1598) you shall labell the houre proposed (vz. 10 of the clocke) sought in the houre circle vnto the Sunne, placed by the 4. Cap. (at 25 degrees 3. min. in ♓) there doth the noone-line shew you (22 degr. 30 min. of ♒) the degree of the-clipticke, beginning the 10 house and the North line (22 degr. 30 min. in ♌) the degree beginning the 4 house, where also (holding fast the Zenitser) play the Almicantifer prepared to the (51 ½) Horizon, on both sides of the noone-line, so shall the pointer cut the clipticke in the East Hemicicle at (28 degr. 30 min. in ♊) his degree beginning the first house commonly called the ascendent or Horoscope, and in the West Hemicicle at (28 degr. 30 min. in ♐) the descendent or degree beginning the 7 house, so haue you the 4 Cardines, Angles, or principall houses of any figure in a manner as ready, as by the particular moouer, which (hauing as the common manner is, drawne your Scheame or Figure for the 12 houses) you shall plant therein accordingly.

But now for the other 8 houses, marke where the Horizontall point (vz. the 51 ½ degr. of the North line) toucheth
on

Astrolabe.

on the Celestiall, and thereto set a marke, or rather the Apex of some loose carect prouided of purpose, (which will happen betweene the two neere feete of Vrsa maior) and there let him rest till anon. And now shall you choose out in the domifying scale annexed to the North line, which slope line serueth vnto your latitude (vz. the 51 ½ slope line) as in the 1 Cap. is touched: vnto whose two extreames or ends, you must seuerally prepare the Zenitser by the domifying rootes, or double radiall degrees numbered in the noone-line scale outwards with Capitall letters in stead of Figures, in all respects as in the 13 Cap. you did for the degrees of altitude: That is to say, setting the Cursor to the like number among the said domifying rootes as the radiall degree, (which also for this purpose is numbered with Capitall letters annexed to the inner ende of the slope line (vz. to the 42 ½ radiall degr. of the North line reckned by the capitall numbers toward the centre) sheweth: wherunto the pointer must be set, if you work for the 2, 8, 6, or 12 houses, or to the outer ende of the slope line furthest frō the centre (vz. to the 57 ½ radiall degr. numbered also by the capitall letters on the outside of the Zenitser) if for the 3, 9, 5, or 11 houses. And first being prepared as is said to y̌ outer end, thē to get the 6 & 12 houses, you shall apply the pointer in the West Hemic. to y̌ Horizon point before left at y̌ carect or Apex (betwene the two neere feete of vrsa maior as is said) & there hold fast the Zenitser, and then playing about the Almicantiser, his pointer shall cut the clipticke on the West Hemicicle, at the (16 de. 30 mi. in ♏) beginning of the 6 house, and in the East Hemicicle at the (16 de. 30 mi. in ♉) degree for the 12 house, set those two into your figure, and then remouing the Zenitser, aply again the pointer in the East Hemicicle, to y̌ same Horizon point at the Apex, & there the Almicantiser still prepared as before plaid about, his pointer shal cut in the cliptick on the West Hemicicle the (20 de. of ♑) degree for the 8 house, & the (20 deg. of ♋) degree Eastwards for the 2 house, then place those two also in the Figure, so haue you 8 of the 12 houses.

Now

The Vranicall

Now for the other 4, you shall new prepare the Zenitfer to the inner end of the slope line seruing the 5, 9, 3, and 11 houses, (vz. by placing the Cursor at the 57 ⅔ domifying roote in the noone-lines scale, and the pointer at the 57 ⅔ radiall domifying degree of the North line, answering the inner end of the 51 ⅔ slope line) and then apply his pointer in the West Hemicicle vnto the foresaid Horizon point left at the Apex, and there holding the Zenitfer fast, and playing about the Almicantifer: his pointer shall cut in the-clipticke on the West Hemicicle, the (5 degree 30 min. of ♒) beginning of the 9 house, and in the East Hemicicle the (5 degree 30 mi. of ♌) beginning the 3 house. That done, then remoue the Zenithfer (30 degrees further) by applying the pointer againe to the sayd Horizon point in the East Hemicicle, & there playing the Almicant. it shal shew you on the East the (22 degr. of ♍) for the 5 house, and on the West the (22 degr. of ♓) degree for the 11 house, now set them all into your scheame, and so is your Figure set.

Fig: Cœli
S:Martij
HOY: 10
1598

Note, that if in playing about the Almicantifer you chance to meete with any notable fixed Starre any thing neere, it shall not do amisse to set him also in your figure. Also I am to note vnto you, that there are two things that fall out vnluckily in this practise, which you shall thus salue. The first is, that the breadth of the Zenitfer hideth from the pointer the Horizon point set at the Apex in the East Hemicicle. This is salued two wayes: the one is, when the Almicantifer is prepared to set the Cursor at so many degrees of the North line.

Astrolabe.

as he is at in the Noone line, and so will the East and West points of the houre-circle change their places for the time: The other way generall to all foure remooues of the Zenith-fer. For so soone as you had gotten the 4 Cardines, if you had made in the Celestialls limbe on each side of the noone-line two markes, the one 30 the other 60 degrees distant from it, that shall serue as well and in stead of the foure times, applying of the pointer to the Horizon point before vsed, as if you trye both wayes you shall finde, and produce one truth.

And whereas for diuers latitudes, some of the domifying rootes do go out of the Zenitfer, those may easily be supplyed with help of a threed and pearle added to the pointer, or with helpe of a paire of compasses thus: First, cast away 45 out of your radiall number shewed by the slope line (vz. out of the $57\frac{2}{3}$ casting 45, there remaineth $12\frac{2}{3}$) and set y Cursor vnto twise so many of the domifying rootes of the noonelines scale as the remaine commeth vnto, (vz. vnto the $25\frac{1}{3}$ domifying roote) and there apply the pointer vnto the like number of the domifying radiall degrees of the North line (vz. vnto $25\frac{1}{3}$ degr. reckoned on the North line from the equinoctials inter-section towards the centre) and then the Cursor remayning, do apply the fiduciall line of the Almicantifer euen with the fiduciall noone-line of y Zenitfer outwards, there doth y poin-ters Aper shew y true place of y roote wanting, where if you pitch one foote of your compasse, or fasten a threed, and with the other or with the pearle extended to the domifying ra-dyall degree of the slope line, (vz. vnto the $57\frac{2}{3}$ radiall domi-fying degree,) offer to describe a circle, it shall cut the-clip-ticke as before the pointer did at 22 degrees of ♓ and of ♍, so the noone-line stand at his due place as before, vz. at 30 or 60 degrees of the Celestialls limbe from the first consti-tution.

♄   Cap.

The Vranicall

Cap. 23.

¶ Of the aspects of the Planets, and how to finde them.

There are in Astronomy three chiefe cardinall aspects (as I may tearme them) vz. the Coniunction marked thus ☌, and is, when as two Planets meete in one degree and minute of longitude. The Opposition marked thus ☍, when as they are iust sixe signes in sunder. The Quartile marked thus ☐, when as they are three signes distant. Also, there are two other of lesse force, vz. the Trine marked thus △, and is at foure signes distaunce. And the Sextile marked thus ✶, at two signes distaunce. Which aspects are easily descryed euen by the very view on this Astrolabe, when as the Planets are thereon planted by the 4 Cap. for who can not espye quickly when any Planet is coniunct, or 2, 3, 4, or 6 signes off from any other, or from any notable fixed Starre. And yet for your better vnderstanding, I haue from the beginning of ♈ drawne the radiall lines of each aspect, and set the Carecters on them: But now Alkabitius Bonatus, and Dariottus will tell you, that the Tryne and Sextile are good, the ☐ and ☍ bad, the coniunction of good Planets good, of euill Planets bad, and that the coniunction and ☍ may be allowed so long as they are within one degree of another, and the △ ☐ and ✶ when they come one within daunger of anothers beames, which for ♄ or ♃ is at 9 degrees before and behinde for ♂ at 8 degrees, ♀ and ☿ at seuen, the ☽ at 12, the ☉ 15 degrees, and so farre off their application and separation beginneth and endeth, &c. Also Io. Stophlerus noteth in this place, propos. 55. that though a Planet might haue 10 degr. of latitude, yet could it not make 30 min. difference in the aspect, and that only in the ✶ or △ and not to be regarded.

Astrolabe.

regarded. Also aspects are said to be derter aspects, when the beames proiect in sequens of the-clipticke, and Sinister when in precedens.

Cap. 29.

¶ A new deuise of the Author to get the Planetary houres in all latitudes, and of their Lords regnant.

Hose houres which in these dayes we call Planetary houres, for that Astronomers haue held opinion the Planets to haue their seuerall dominions in euery of them, haue beene sometime called Horæ inequales, and Horæ temporariæ. Their nature is to diuide euery naturall day betweene Sunne-rising and setting into 12 equall partes, and euery naturall night into other 12. But because in all oblique Horizons the dayes and nightes lengthen and shorten with the Sommer and Winter, therefore these houres growing great and lesse according to the time and latitude, are called Inequales and Temporariæ, they are numbered in the day time from Sunne-rising, ending at Sunne-set with 12, and there the houres of the night beginning, do end the next morrow at Sunne-rising with 12.

These houres are most easily had by a speciall Instrument called Horarium Planetarium, which I my selfe haue newly deuised, generall to all Horizons, which was neuer done before me that I haue seene, and placed it at the top of the Celestiall, comprehended in the Sertans of a circle, whose arche contayneth the houres of Sunne-rising and setting, and within, it is filled with a number of streight lines, all parallell to the base of the Sertans, representing the common, equall, and vsuall houres either of day or night, which are numbered on the Sertans side downewards to 12, for the houres going before the sirt Planetary houre, and thense backe

H 2 againe

The Vranicall

againe outwards for the houres after. And it must haue riding on his Centre a Labell or Scale of the Planetary houres of equall length to the side or base, deuided into sire equall parts, numbered from the outer end inwards, ending at the Centre with 6, and thense backe againe, ending at 12 where it began. This Labell or Index is also numbered from the Centre outwards vnto 12 for the vse of the 7 Cap. Now when the common houre of any day or night is giuen, (admit 10 of clocke 40 minutes afternoone the 12 of December 1583, being the houre of the natiuity of my neere kinseman Maister Walter Seint-Iohn, sonne and heyre to Sir Iohn Seint-Iohn Knight of Lyddiard in the County of Willshire deceased) and that you would know the Planetary houre, you shall by the 17 Cap. first get the houre and minute of $\breve{y}$ Sunne setting that day, if the houre giuen be of the day or of the Sunne rising, if the question (as that natiuity was) be of the night (vz. you shall finde it about $8\frac{1}{4}$ of clocke in the morning) and vnto that houre (vz. $8\frac{1}{4}$) sought in the Sextans arche, set the fiduciall line of the Planetary Scale, where immediately you shall see the houre-line equall to your houre giuen (vz. the $10\frac{2}{3}$ houre-line reckoned in the Sextans side inwards for that midnight will be the 6 Planetary houre) to cut off in the Planetary Scale the Planetary houre desired. (vz. it shall cut $\frac{2}{3}$ partes of the 9 Planetary houre.) And if now you desire to knowe what Planet beareth dominion that houre, resort vnto the domineering Table which I haue placed at the lower corner of the Celestiall, and therein seeke the day or night of the weeke proposed, accounting Sunday for the first, Munday the second, &c. wherein your houre was giuen. (vz. Thursday night being the 5 night I seeke the figure 5 by the title (Noctes) and follow that range of Carects vntill you come vnder your Planetary houre founde, (vz. vnder the houre figure of 9) and there shall you finde ♄ the Planet domineering that houre.

Note, that if you had no domineering Table, you might easily

Astrolabe.

easily make it by hart, beginning the first houre on Sunday with the ☉, and then reckoning the Planets on your fingers ends thus, ☉, ♀, ☿, ☾, ♄, ♃, ♂, ☉, ♀, ☿, ☾, ♄, &c. till you haue done 24 so shall you finde the 25 in this accoumpt to be ☾, and thereof Munday is called dies Lune, and on Tewsday will fall out ♂ if you reckon on, and thereof called dies Martis, euen so Wensday, dies Mercurij, of ☿, Thursday dies Iouis, of ♃, Friday dies Veneris, of ♀, Saterday, dies Saturni, of ♄, and all by reason of this accoumpt of the 7 Planets.

Cap. 30.

¶ How you shall knowe to set forwards and backwards any of the fixed Starres vnto their true places for any age past, or to come.

I haue in the 2 Cap. sufficiently shewed by what meanes the fixed Starres are in apparance from age to age, forced out of one signe and degree into another, through the sliding backe of the Equinoctiall intersections with the clipticke: proceeding orderly and equally almost euery 67 yeares a degree, which is called the Medius, or Æqualis motus precessionis seu fixarum. The gate whereof, how much it is in one day, in one yeare, in 4, in 20, in an 100, and in a 1000 yeares, is set downe in the first columne of the Tabula Radicalis, in the left corner of the Celestiall. Wherefore if you would knowe where the equall place of any Starre (admit of Arcturus) shall be any time to come, (admit for the yeare after Christ 2000 compleate) take 1650 (for that all the North Starres are in this Celestiall placed and rectifyed for that yeare) out of the yeare proposed, and by that first columne gather the equall motion aunswerable to that remayne, (vz. 1650 taken out of 2000 leaueth

The Vranicall

350, for which 350 yeares, I take in the said columne the motion of one 100 yeares three times, and of 20 yeares twise, and of 4 yeares twise, and of one yeare twise, for that three times 100, and twise 20, and twise 4, and twise one, do make 350, and of that collection there amounteth 4 deg. 53 min. almoſt.) And ſo much can I ſay that thoſe North Starres will then be gone forwards from their places, where now in the Celeſtiall they are ſet in regard of their equall motion. But now there is found out by ſkilfull artes-men a certaine libzament or ballaicing of the Equinoctiall and the Poles of the world, as the ſtudyed in the Theoricks well knowe, which at one time helpeth forwards this equall motion, and at another time pulleth it backe againe, euen vnto 71 min. or 1 degr. 11 min. all is one, making by that meanes the true motion vnequall. Therefore I haue within the Radicall table placed an inſtrument of my deuiſe to equate the ſame, which conſiſteth (as the Radicall table doth) of two partes, the one for the ablatiue Æquations, the other for the Adiectiue, each part hauing two limbes ioyning together, the outermoſt carrying the ſayd 71 minutes two wayes for the ballaicing too and fro: the innermoſt being double, in which are written the yeares from Chriſtes birth for 3440 yeares after, numbered from the ✶ twiſe about, for that in ſo many yeares the Anomalye of this ballaicings compleateth twiſe, & therefore that inner limbe is double. Wherefore if now you would knowe this Æquacion for any yeare (admit the yeare 2000) ſeeke the ſame yeare 2000 in that double limbe, & you ſhall finde in the outer limbe the minutes of equation that anſwere thereunto (vz. almoſt 70 min. ablatiue) to be added or ſubducted too or from the equall motion before gotten accozding as the words Adiectiuæ or Ablatiuæ do admoniſh. (vz. whereas the equall motion of the fixed Starres was before found for the yeare after Chriſt 2000 compleate to be gone forwards 4 deg. 53 mi. you muſt now pull out 70 minutes, ſo will there be left 3 deg. 43 min. their true motion in that 530 yeares, according to

which,

Astrolabe.

which, the true place of Arcturus, which for the yeare 1650 is placed in this Celestiall 18 degr. 55 min. in ♎, shall then be 22 degr. 38 min in ♎. And so for any other North Starre, for they all remooue alike.) And likewise for any of the South Starres, but that you must for them worke vpon the yeare 1570, as before you did by 1650. Also for any age past (as if it had beene for 350 yeares before 1650. vz. the yeare after Christ 1300 compleate) you should haue found the Equation as much adiectiue, and by that meanes the true motion of 350 yeares backwards 6 degr. 3 min. to be pulled backe from their now place, (so that Arcturus was then but 12 degrees 52 min. in ♎.

Cap. 31.

¶ How by help of this Astrolabe to get the longitude and latitude of any Planet, yea, or of any Starre or Comet seene.

Hauing a Meridian exactly set as in the 9. Cap. is taught, you shall obserue the comming of any Planet, Starre, or Comet to the Meridian either in the South or North partes, (as for example the 2 of October 1596 I obserued the comming of ♃ to our Meridian heere at Reading) and there take his meridian altitude (which I found 50½ deg.) as is taught in the 11 Cap. and presently (except you haue some clocke or watch truly set) get the houre and minute if you can by taking the altitude of some knowne Starre, as in the 13 Cap. is taught, (I found it a quarter of an houre past one and better after midnight) which done, seeke that houre and minute in the houre circle, and labell it to the ☉ planted by the 4 Cap. (at 20 de. 10 min. of ♎) & there hold fast the Zenitfer, & number the verticall distance or complement of the Merid. alti. before taken (vz. 39½ de.) on ȳ nooneline frō ȳ Zenith point towards Meridies, if your Merid. alti. taken were South, or towards

Media

The Uranicall

Media nox, if North, (which heere being South, will fall out at the 12¼ radiall degree numbered from the Æquinoctiall) where make a pricke on the Celestiall, for there shall be the true point or place of the (♃) Starre, Planet, or Comet, obserued for that instant, and the circles of the-clipticks longitude and latitude cutting that point, shall shew you his longitude and latitude sought (vz. the circle of longitude passing by that verticall distance or pricke made, shewed me in the clipticke 8 degrees 30 minutes almost of ♉, the longitude of ♃, and the circle of latitude cutting that pricke, shewed me almost 2 degrees Southwards from the-clipticke his latitude.) And this serueth for the ☾, or any Planet, Starre, or Comet seene. And now by helpe of this latitude thus gotten, you may as easily get the true places of the Starres seene in any other place as well as in the Meridian perpetually, for that their latitudes neuer alter, and of ♄, ♃, or ♂, 20 or 30 dayes after: for that theirs altereth little in that time, and of ♀ for a day or two after, but not for the ☾, because her latitude changeth swiftly, neither for ☿, because in our Clymat we neuer see him: In this maner.

Take the altitude of any Starre, Planet, or Comet, whose latitude is foreknowne either by this Cap. or otherwise at any houre and minute knowne, taking good heede whether he be in the East or West Hemispheare. Then prepare the Zenitfer to that altitude taken, then labell the houre and minute to the ☉, and there hold fast the Zenitfer. Which done, apply the pointer to the circle of latitude of the Starre or Planet either in the East or West Hemicicle, according as the Starre or Planet was at taking his altitude, so shall the circle of longitude passing by the pointer so applyed, shew the longitude of the Starre or Planet in the-clipticke.

Note, that Io. Stophlerus lib. 2. prop. 38. reciteth 7 propositions which some lying Astronomers (as he tearmeth them) haue taken vpon them to teach on the Astrolabe, which he there condemneth to be most false, and not possible, and therefore

Astrolabe.

fore spendeth a Chapter to forewarne euery man to take heede and not to meddle with them. And these are the 7, the first, to knowe the longitude of the ☽, the second, the longitude of ♄, ♃, ♂, ♀, or ☿, the third, to get their latitudes, the fourth, to know whether a Planet be direct or retrograde, the fift, to know the longitude of any Starre in the Reete, the sixt, the latitude of fixed Starres, the seuenth, the longitude of Starres not in the Reete. And now I am to rewarne all men, that they are all possible by this Astrolabe, and by my Iewell also, and the whole 7 sauing the 4, are performed by this one Chapter: because I would not haue them now daunted with that 38 propos. of Stophler who meant it no doubt only, by that Astrolabe whereof himselfe wrote.

Cap. 32.

¶ A ready way by helpe of this Astrolabe, heereafter to obserue the motions of Comets.

Ecause I haue seene some 4 or fiue yeares past a booke entituled Noua theoria Cometarum as I remember, set foorth by one Reslyn, who taking occasion vppon that great Comet or Blazing-Starre, which Anno 1570 was seene so long in the Constellation of Cassiopeia, in a manner fixed without motion, to imagining therefore that Comet to happen in the very Pole of the Theoricke, and that to be the cause why he mooued not, thereupon runneth on a course, with recitall of diuers Comets and their motions, but concluded no certaintie to my remembrance, ending his booke with this saying, Est quodam prodire tenus si non datur vltra. But our late learned countreyman

The Vranicall

countreyman, Mayster Digges, in his Scala Mathematica found, because he had no Parallax, that he must needes be beyond the Speere of the ☾. Well no doubt there can not any weighty thing come to perfection at the first, second, or third assay, and therefore to the end it may be a generall care to obserue the motions of Comets when they happen, I haue thought good to aduertise euery man in this Cap. that either on this Astrolabe or on the Globe, it is a most easy matter when any Planet appeareth, to pricke downe his place euen by the very view of his scituation among the fixed Starres, seeing by the 25 Cap. before you may so easily learne to know them all.

As for example in July this yeare 1596 there appeared a Comet with a streaming tayle vpwards in the North-west, which vpon Teusday the 20 of July aforesayd, about 9 of the clocke at night, I first tooke heede of betweene the two hinder legs of Vrsa maior almost in a streight line, (but somewhat lower, and two parts of the way) betweene the Starre of the fourth light in the neere hocke of the neere hinder legge, and the two Starres of the third light in the further foote behinde, according whereunto I made a pricke in the Celestiall of this Astrolabe. The 21 of July the same houre, I sawe him directly in the streight line aforesayde, and distant from the sayd two Starres twise their distance. The 22 day the same houre I sawe him a little aboue the streight line, and but their distance off. The 23 of July the night was darke. The 24 of July I sawe him aboue those two Starres of the third light, and to make an equilater triangle with them : all which, I noted downe euery night with pricks, then followed a fewe cloudy and darke nights, and after that 24 day he was no more seene. This Comet was seene of many others some fortnight before I sawe him, as neere as I could gesse by their relation euen about the brisket of Vrsa maior, a little behinde the hocke of the neere legge afore, which also I pricked downe : and haue caused

those

Astrolabe.

those prickes and the trayle of his gate to be grauen in the Celestiall, in the Const:llation of Vrsa maior if you marke it, (though not done fully to my purpose) to the end to incite others heereafter to do the like, that posterity at the last may finde out whether there be any regularity in their motions. If the Comet chaunce to come to the Meridian either in the North or South (as this did not in sight) then by the last Cap. you may get his longitude & latitude euery night and so pricke downe euery dayes motion the more exactly.

FINIS.

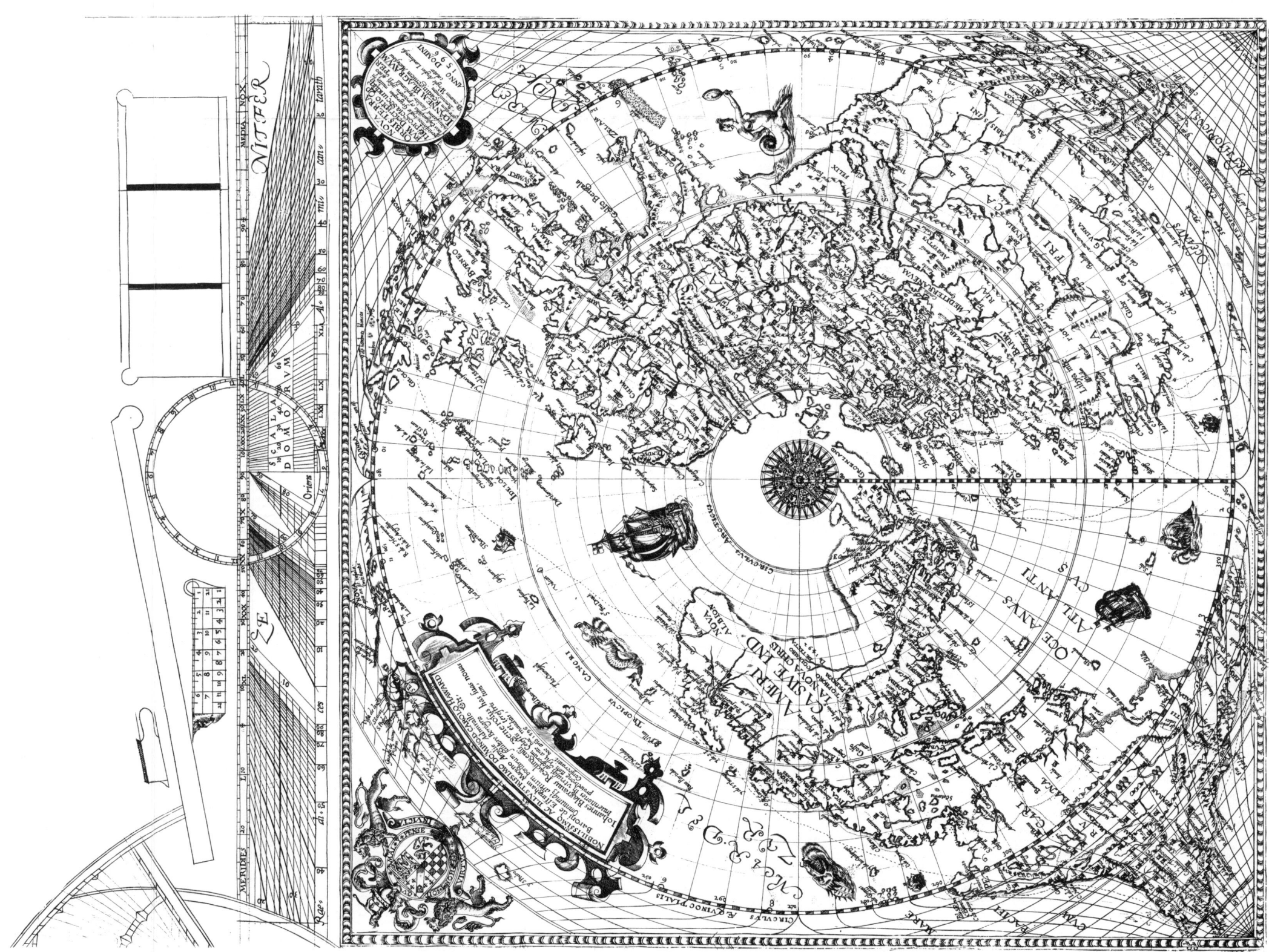

QB
85
B6
1972

DEC 5 1973